Hello Meth Lab in the Sun

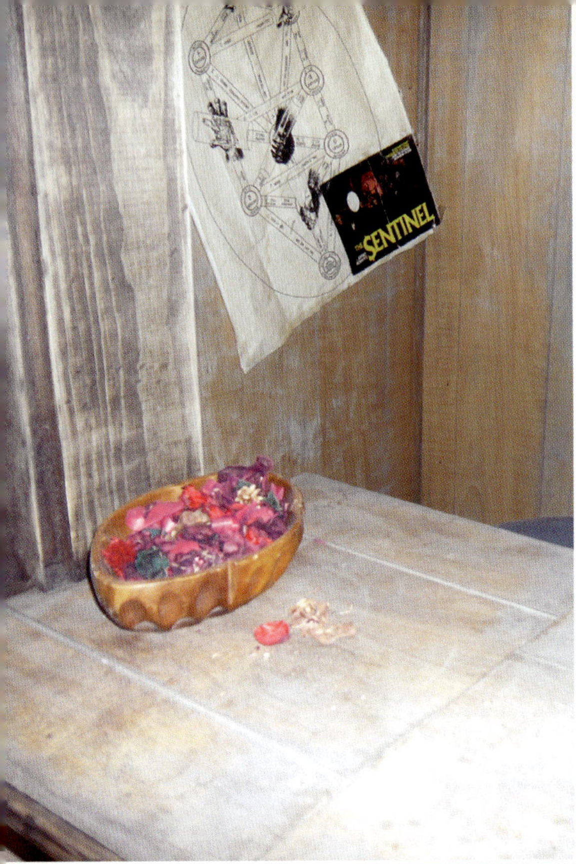
SENTINEL

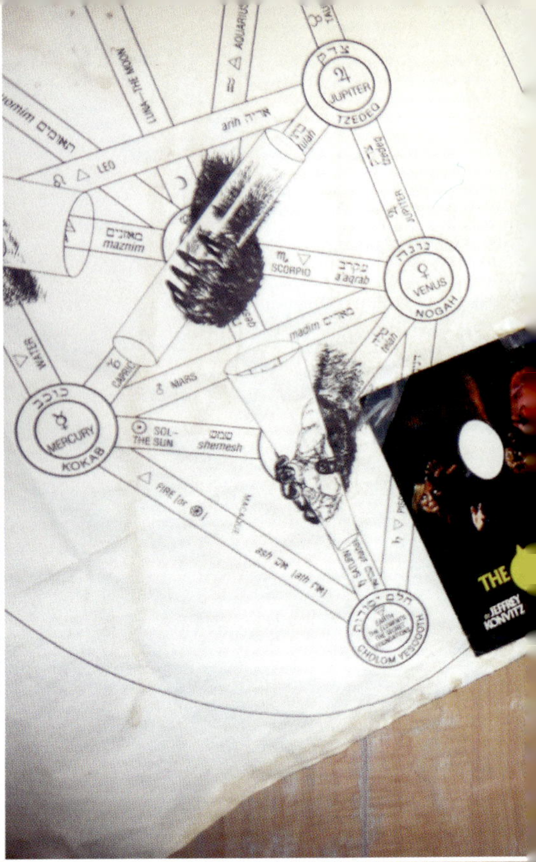
JUPITER
TZEDEQ
VENUS
NOGAH
MERCURY
KOKAB
LEO
SCORPIO
a'aqrab
MARS
SOL-
THE SUN
shemesh
mazmim
madim
WATER
THE
JEFFREY
KONVITZ

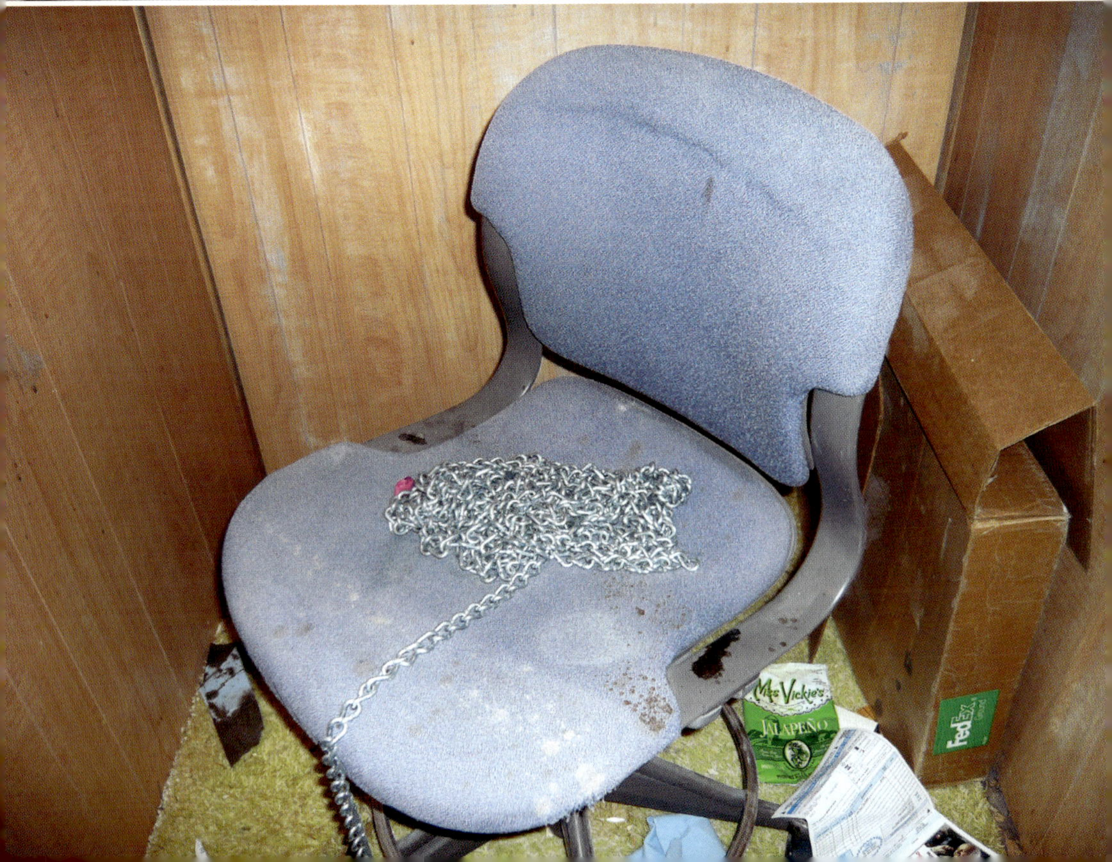
Miss Vickie's
JALAPEÑO
FedEx

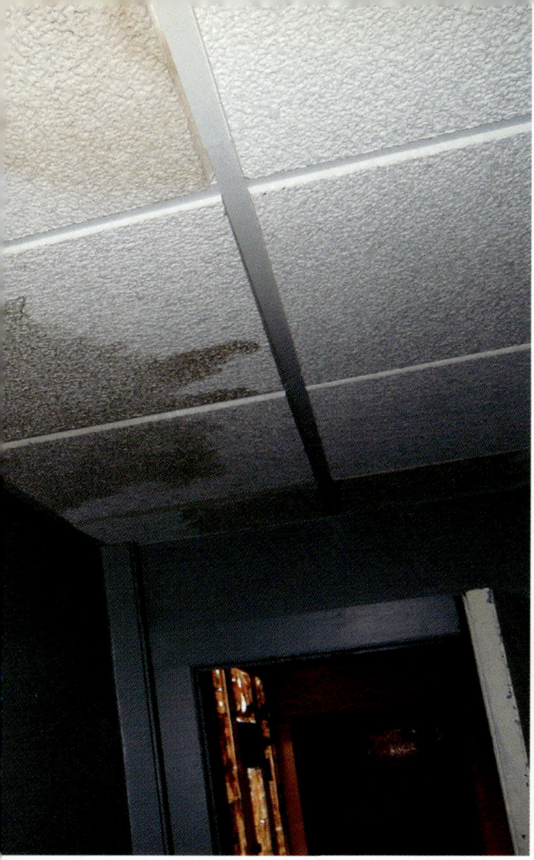
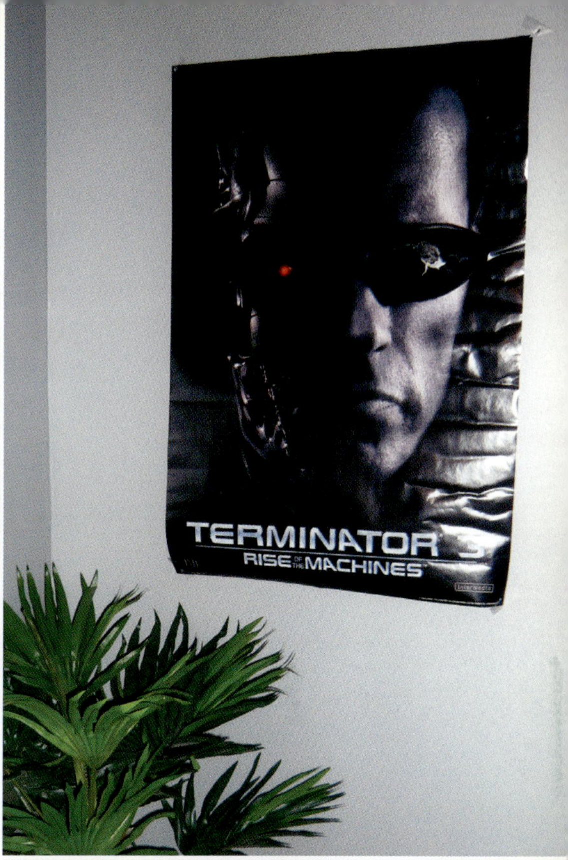
TERMINATOR 3
RISE OF THE MACHINES
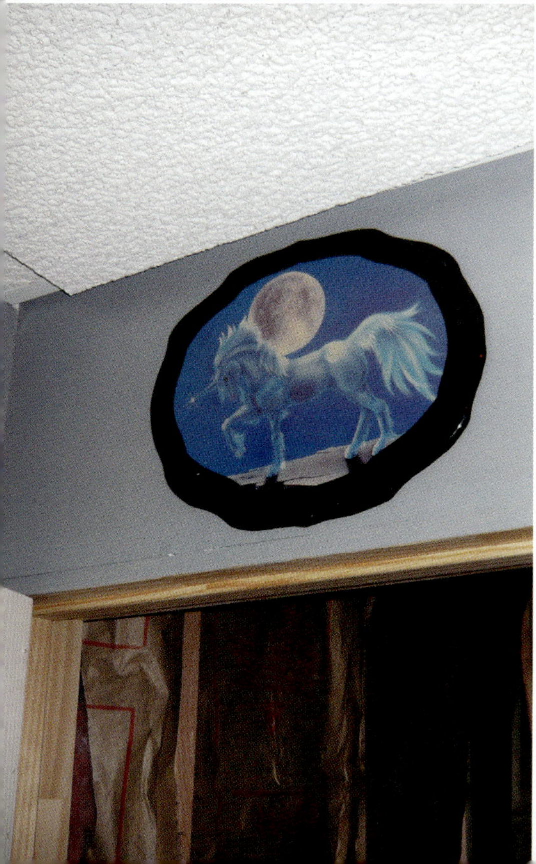
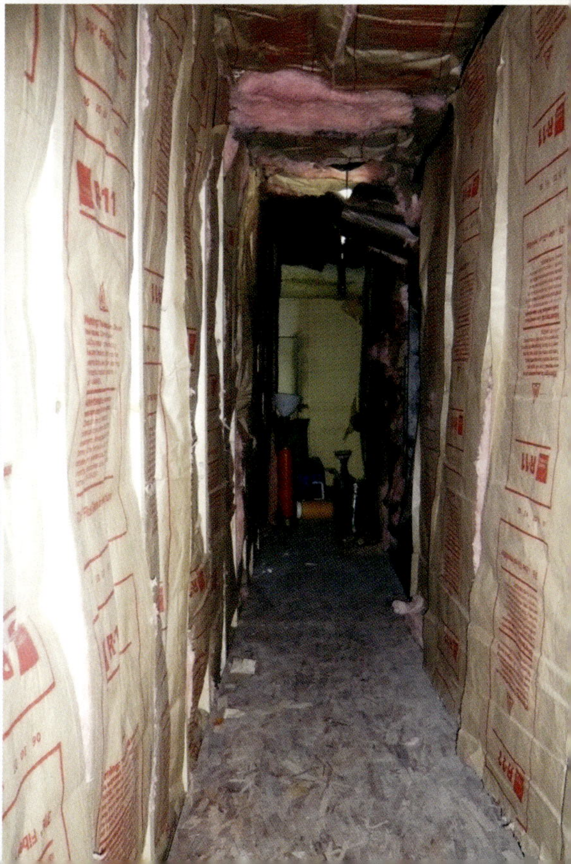
R-11

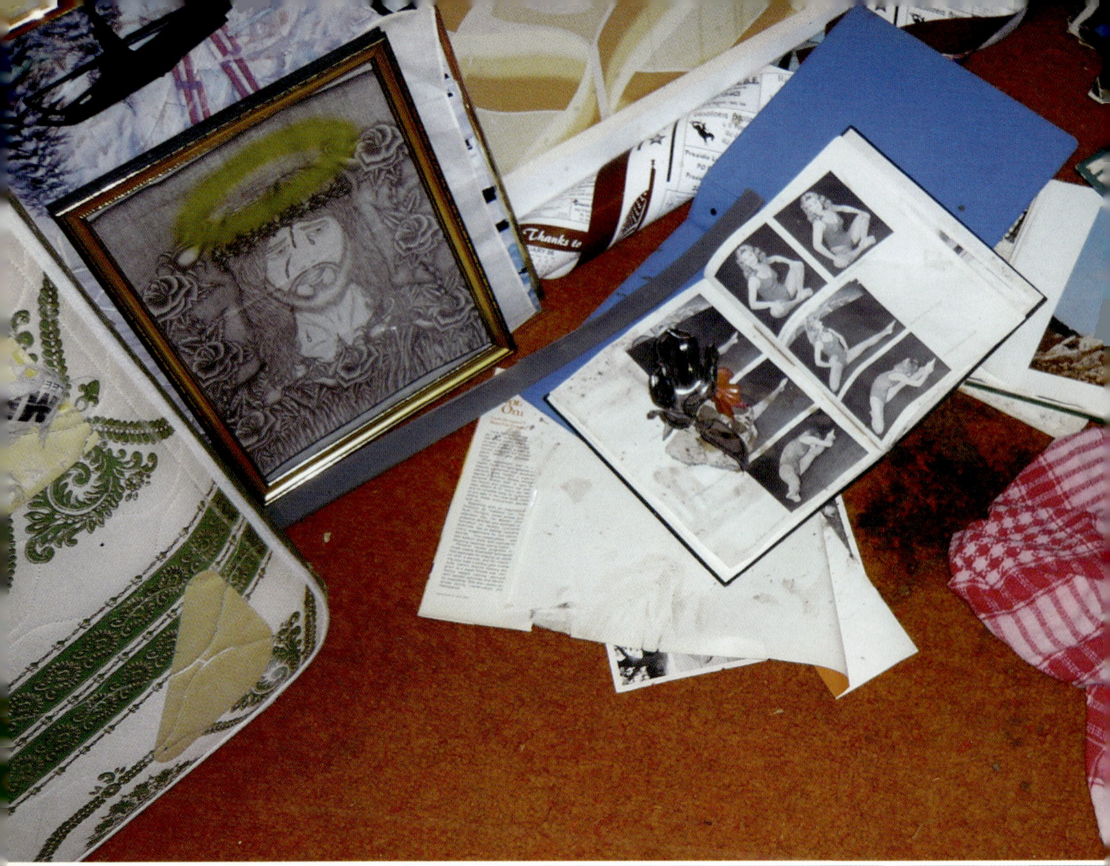

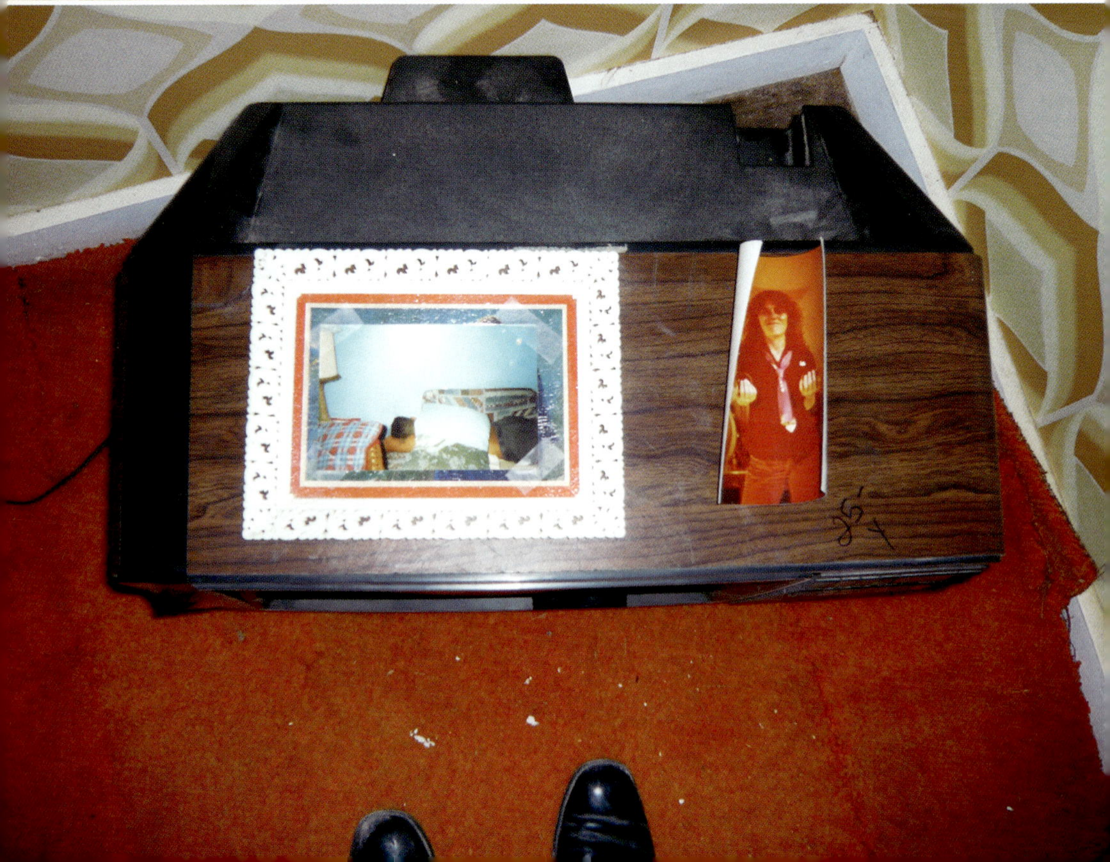

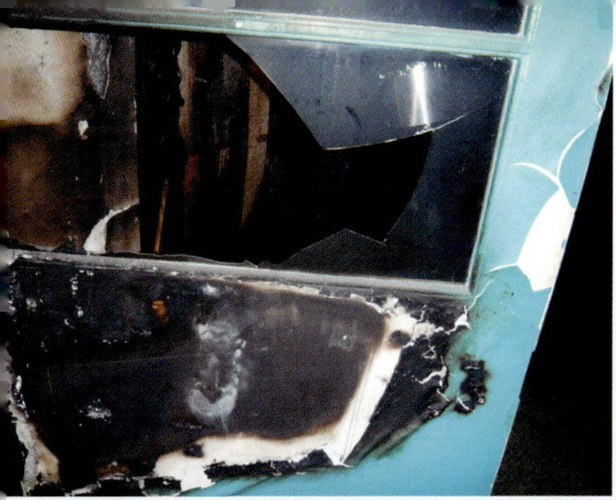
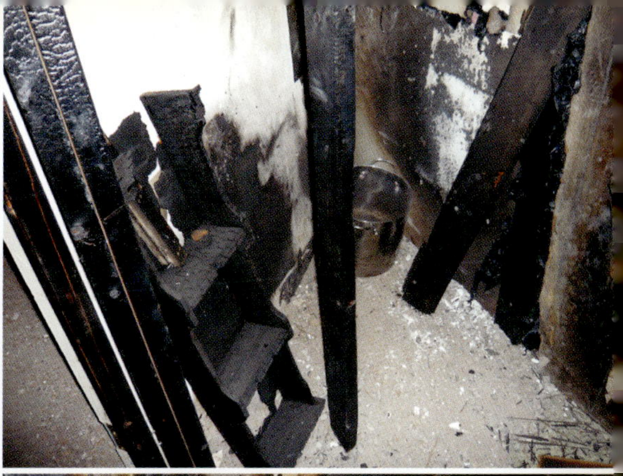
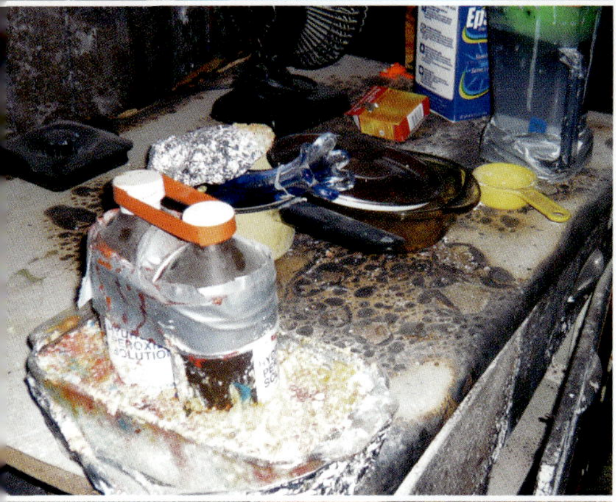
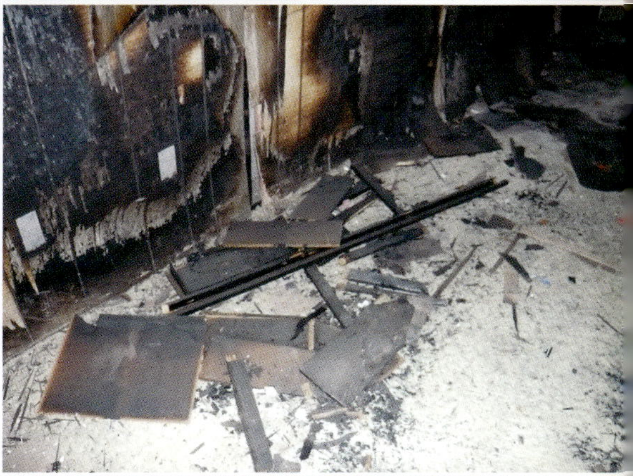
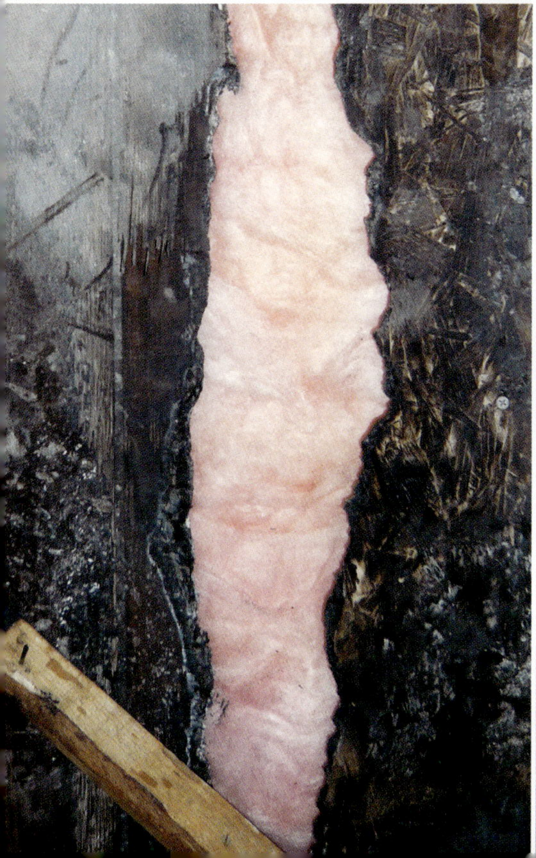
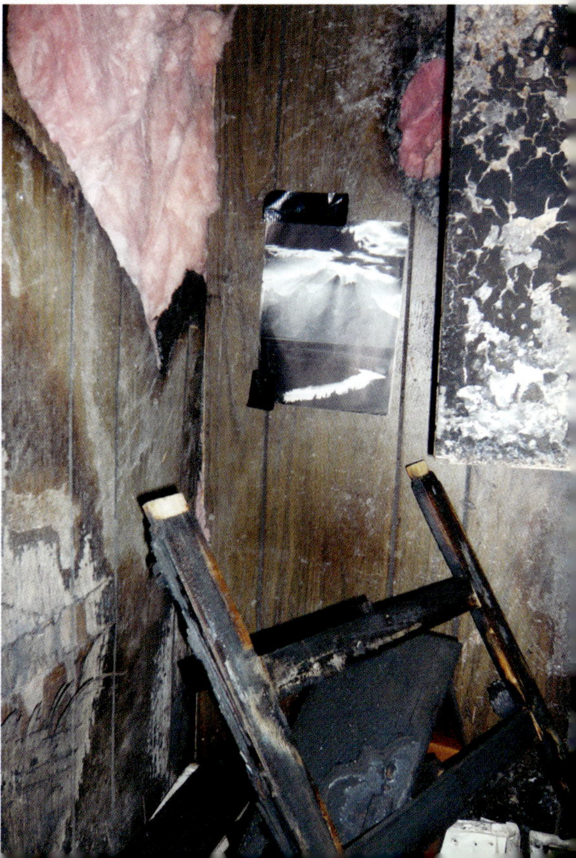

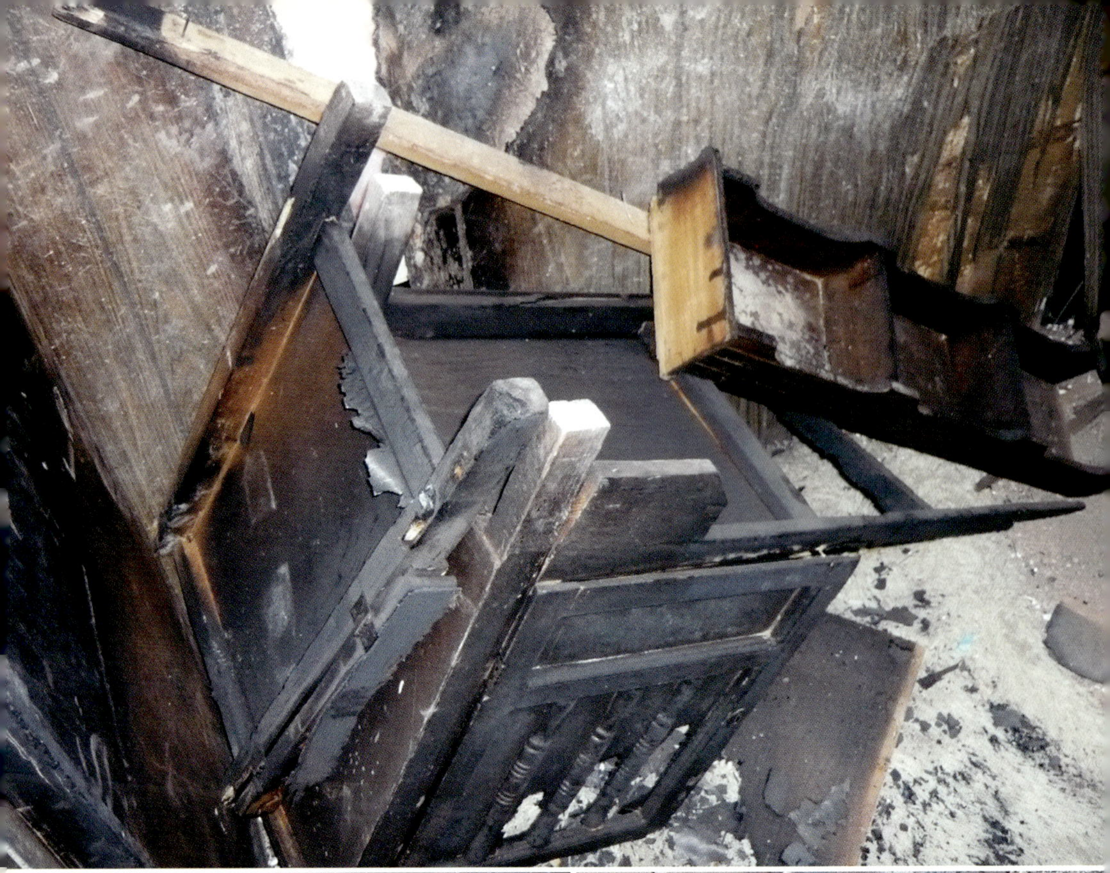

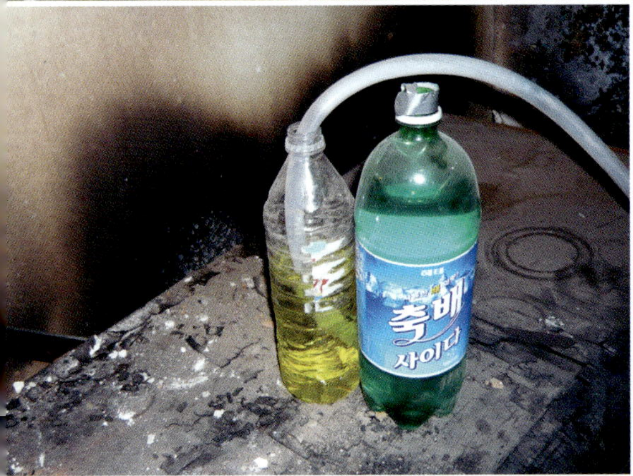

축배
사이다

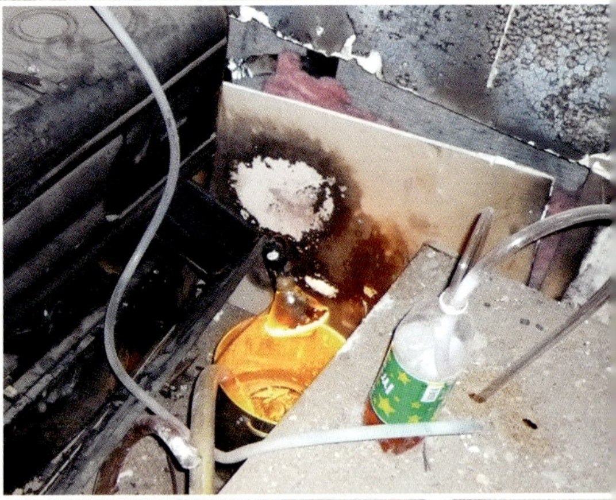

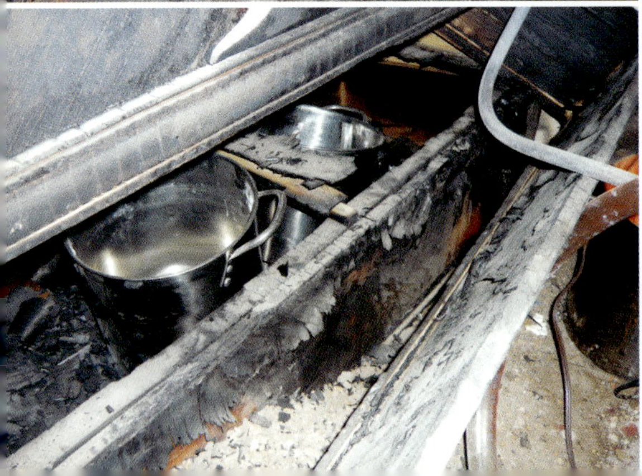

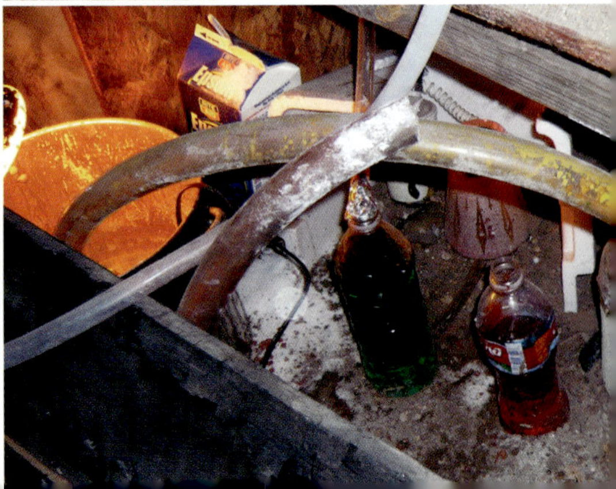

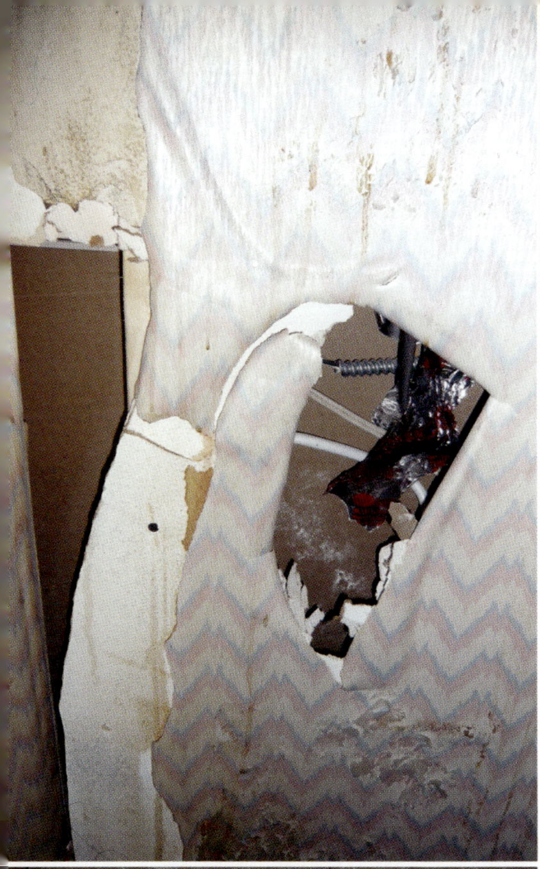

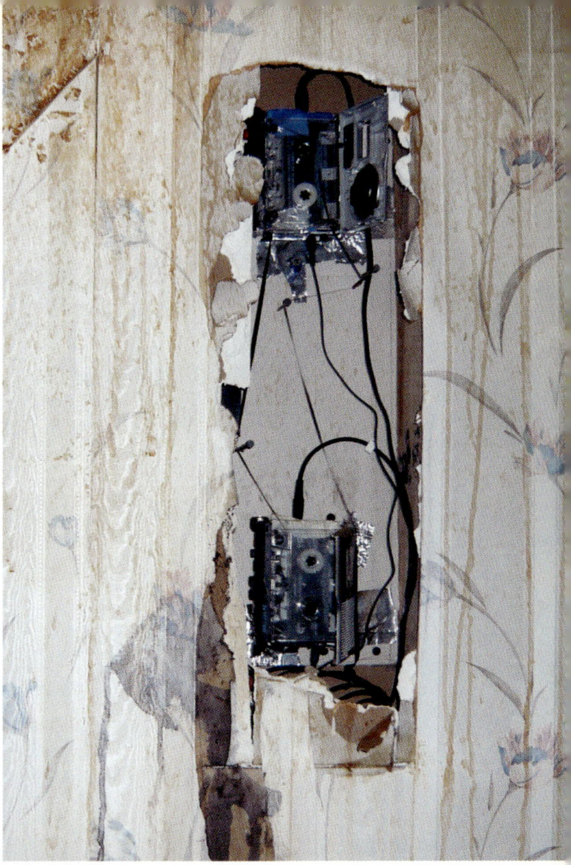

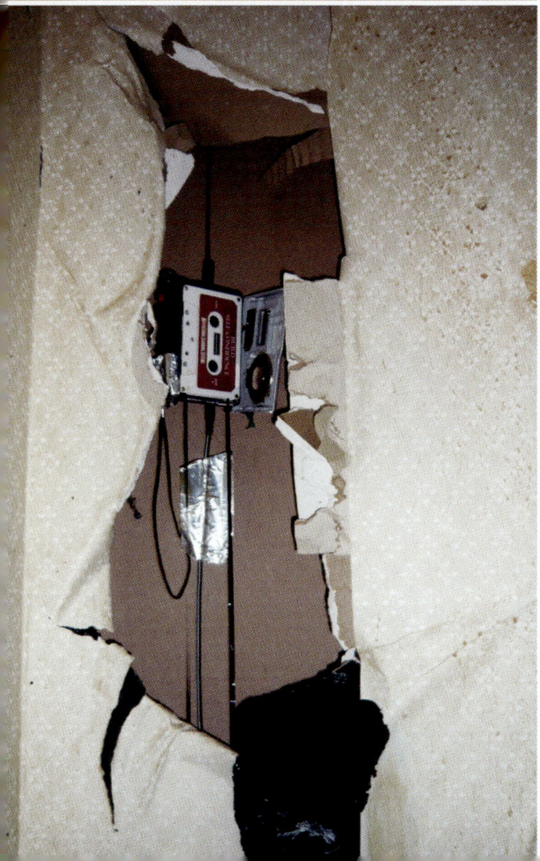

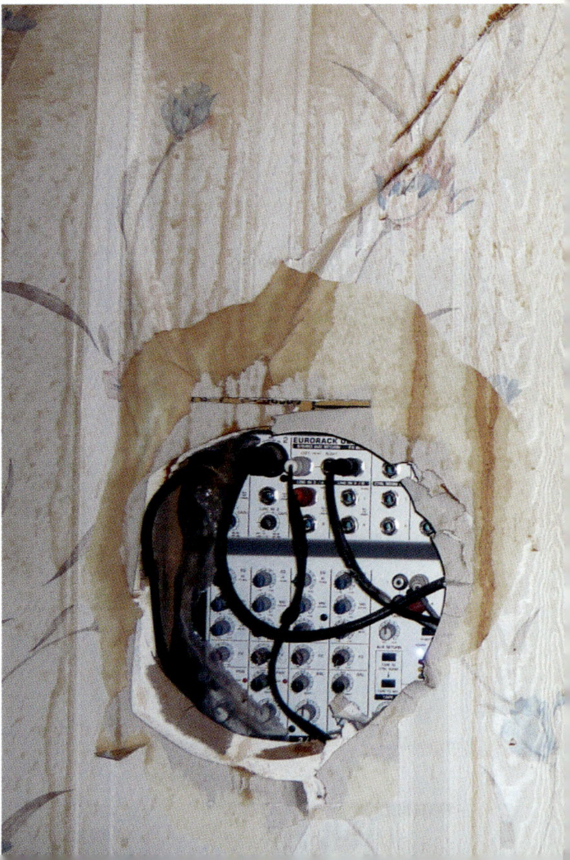
EURORACK

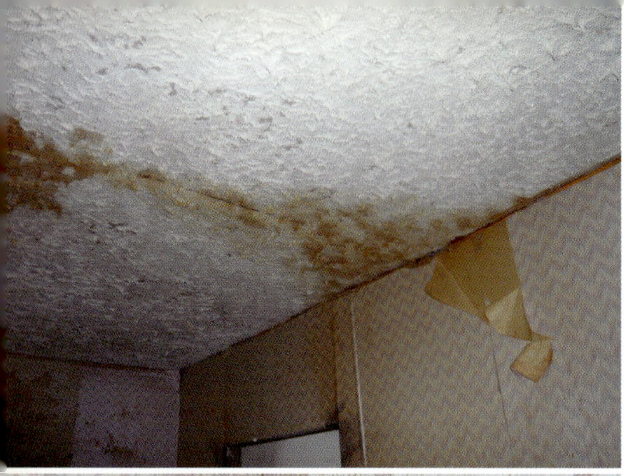
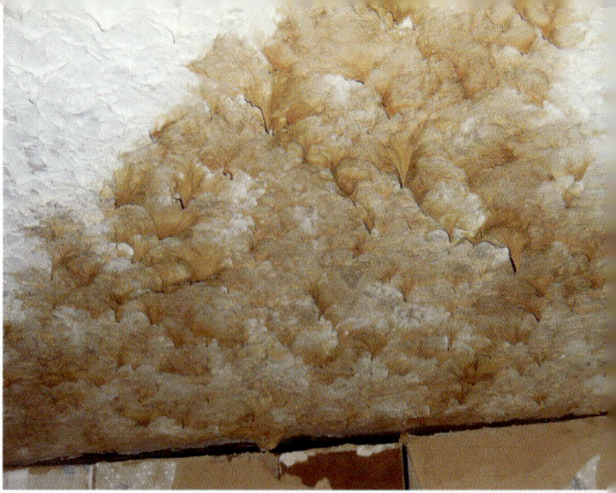
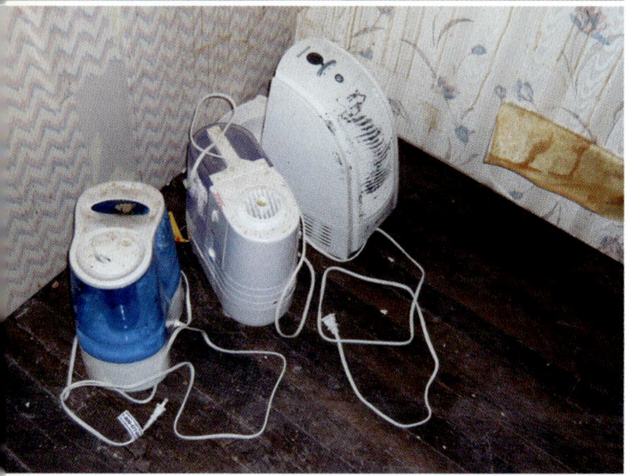
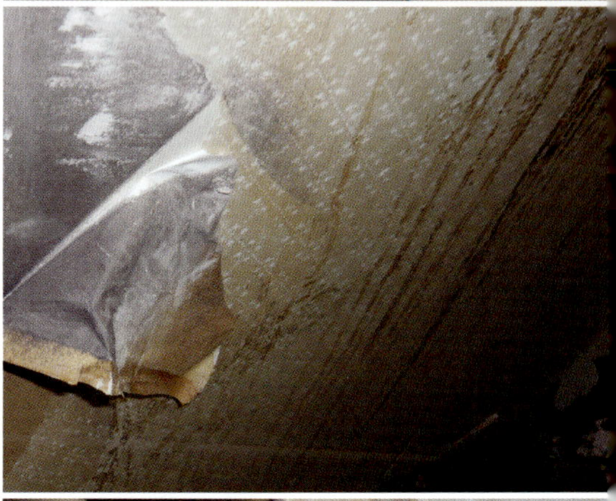
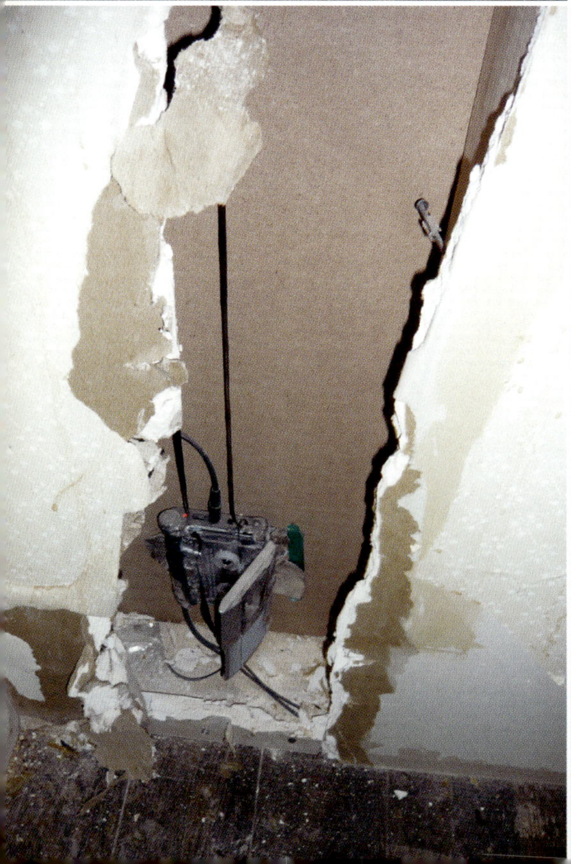
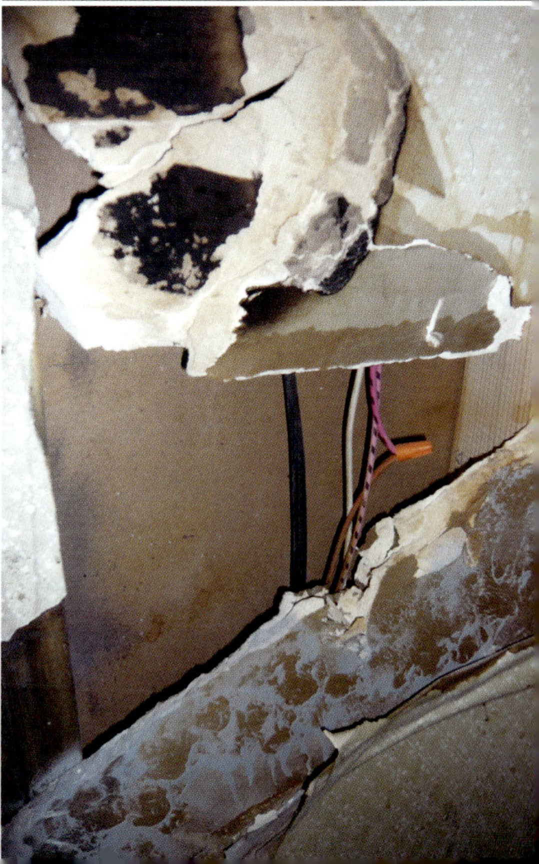

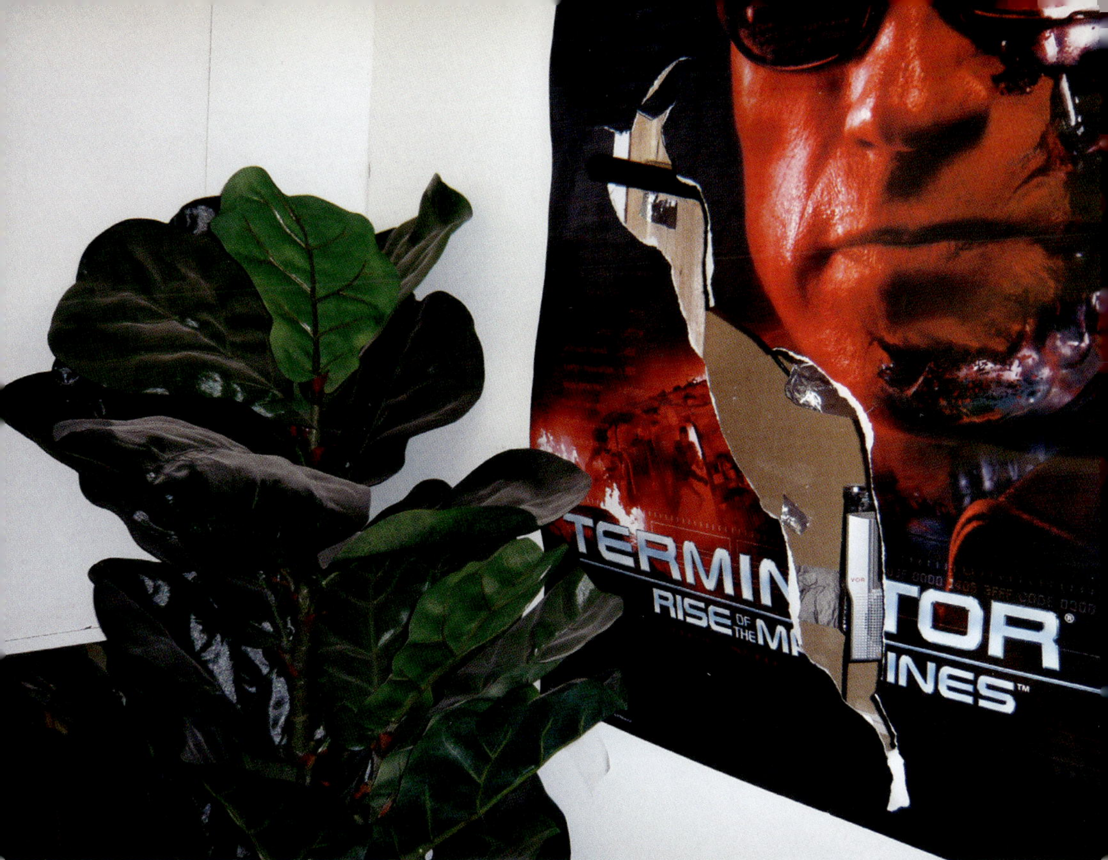
TERMIN TOR®
RISE OF THE MA INES™

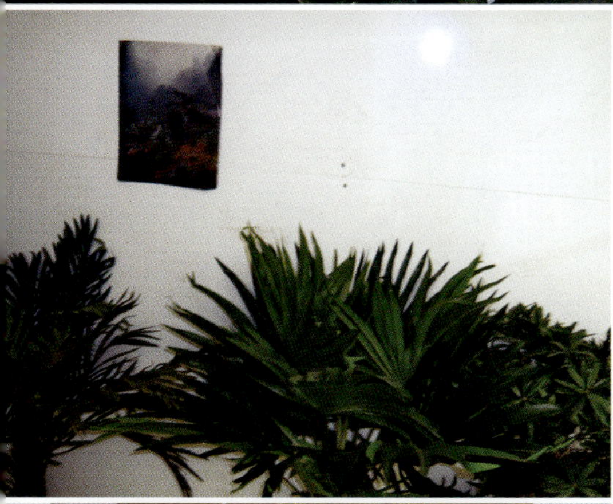

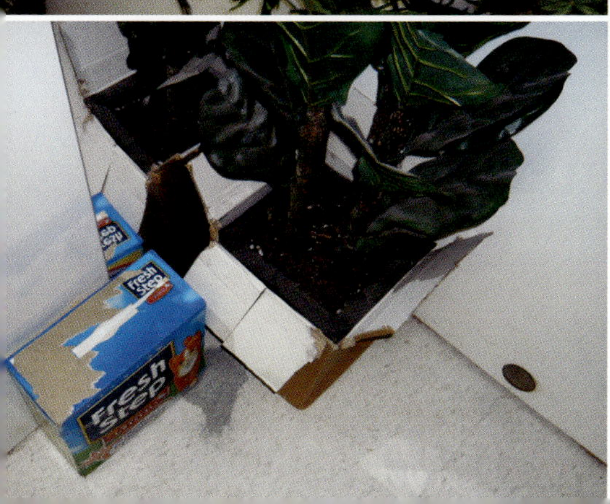
Fresh Step
Fresh Step

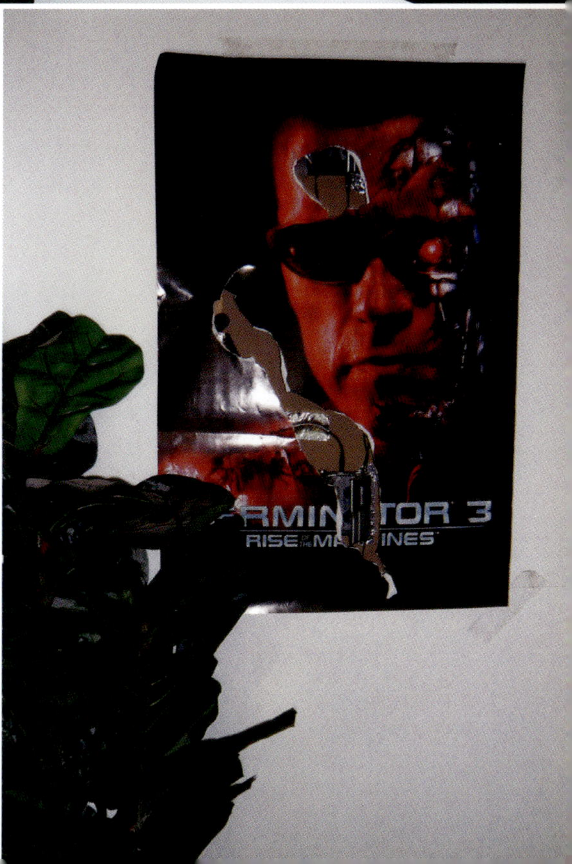
RMIN TOR 3
RISE OF THE MA INES

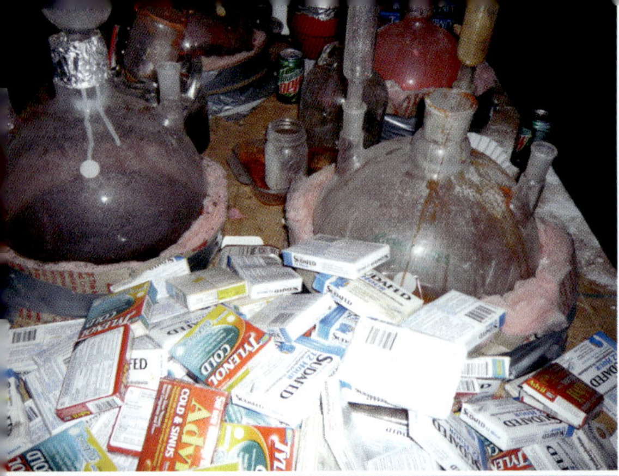
TYLENOL COLD

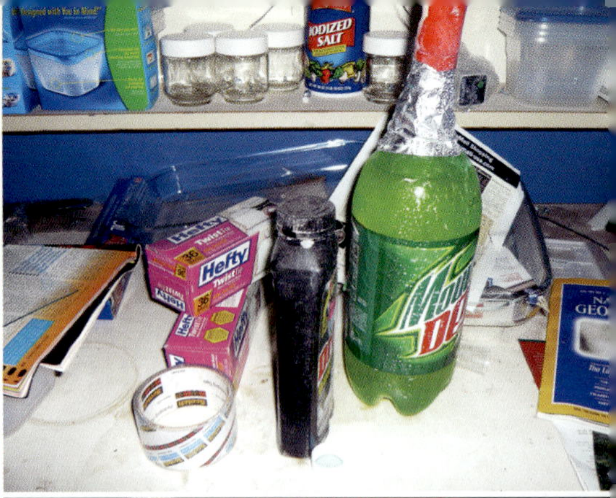
IODIZED SALT
Hefty

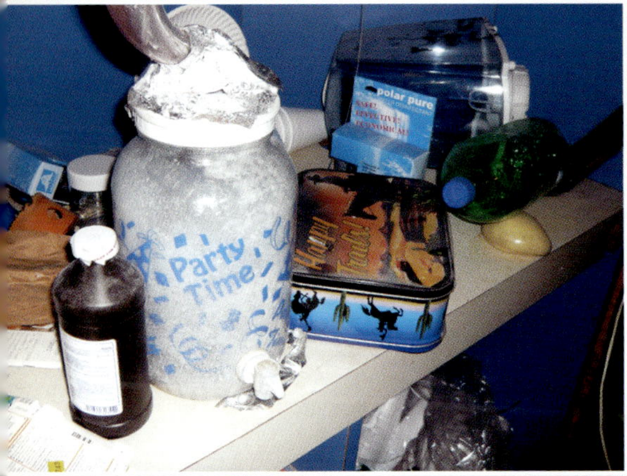
Party Time
polar pure

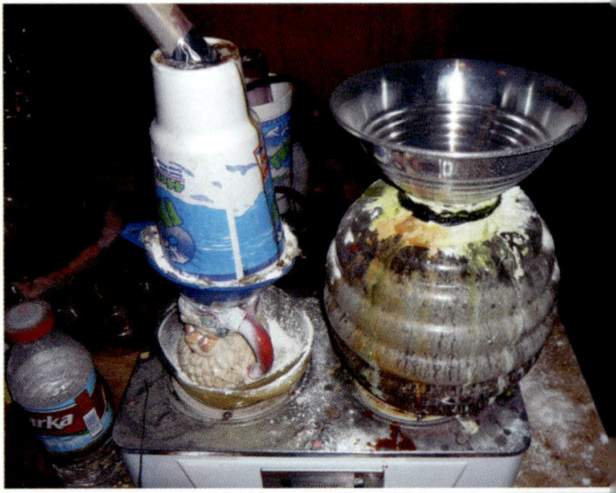

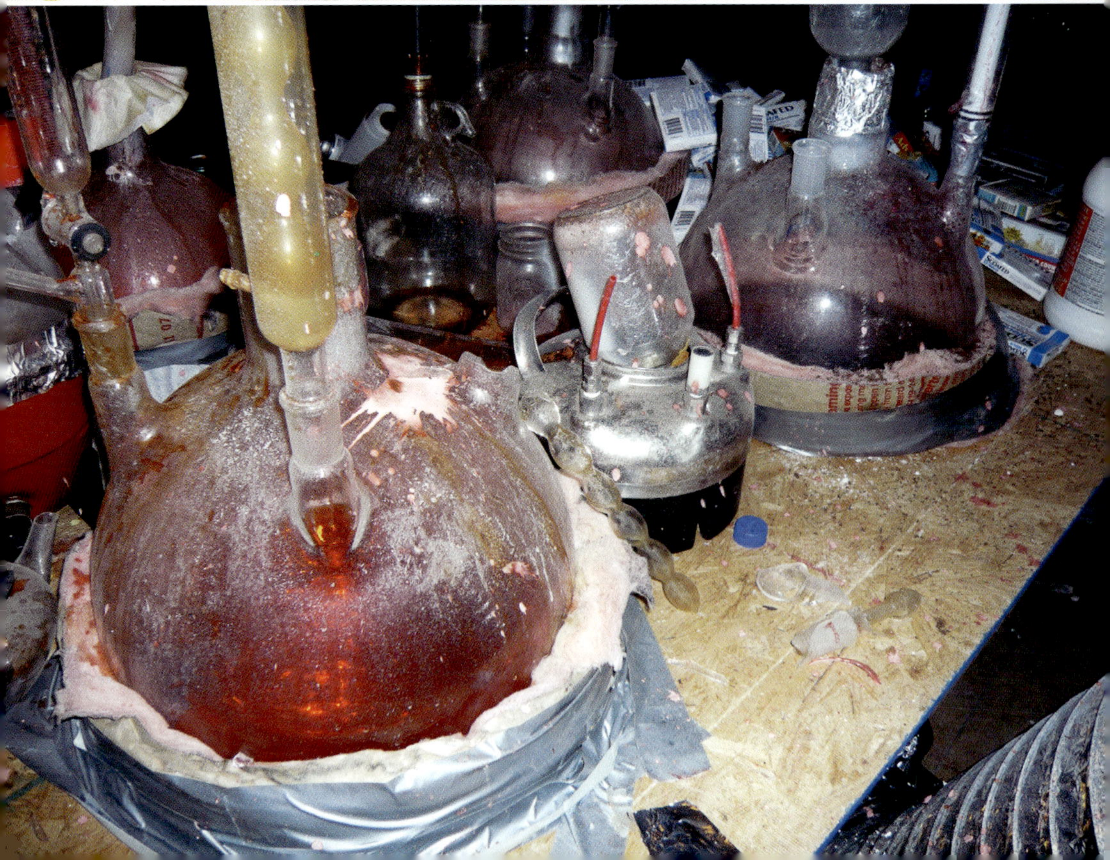

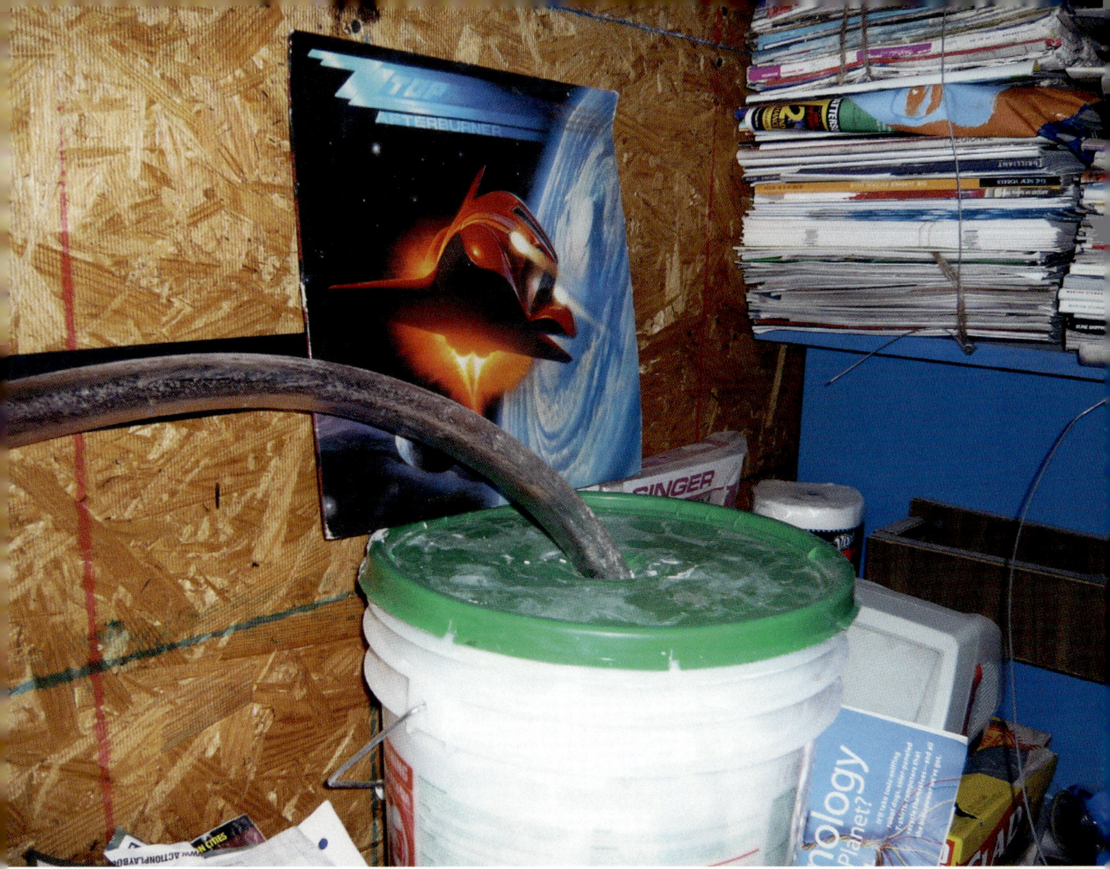
AFTERBURNER
SINGER
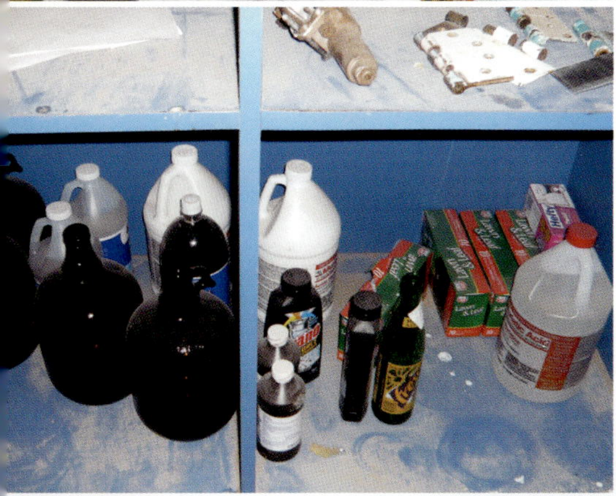
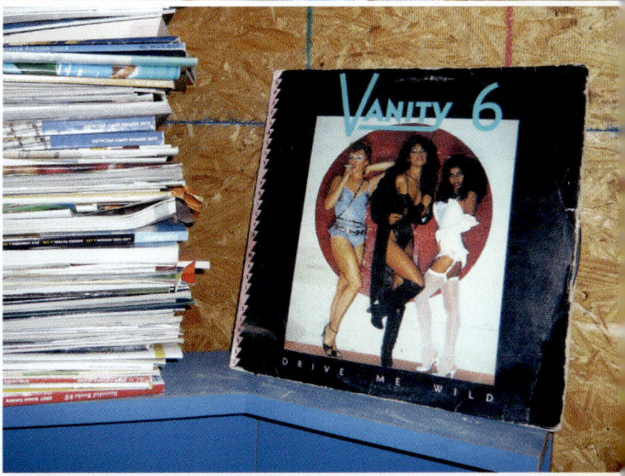
VANITY 6
DRIVE ME WILD
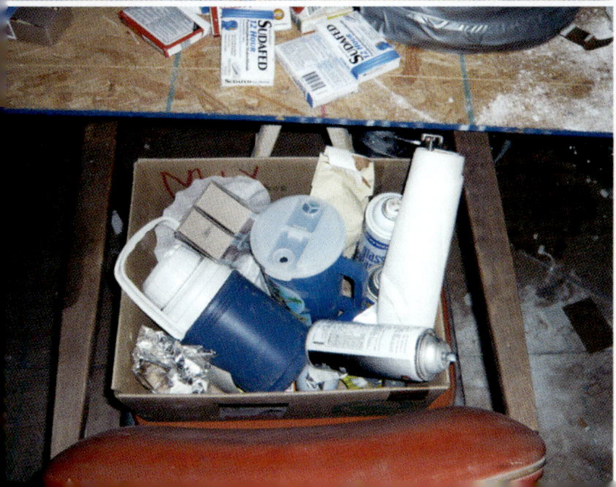
SUDAFED
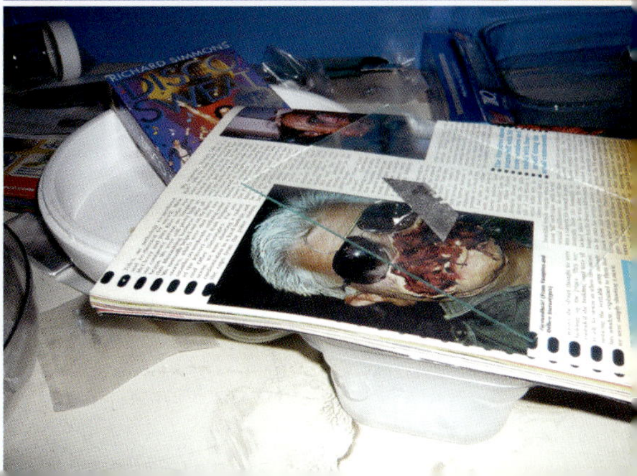
RICHARD SIMMONS

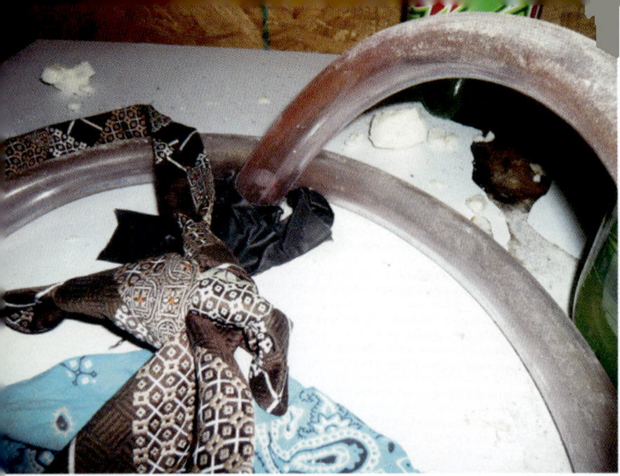
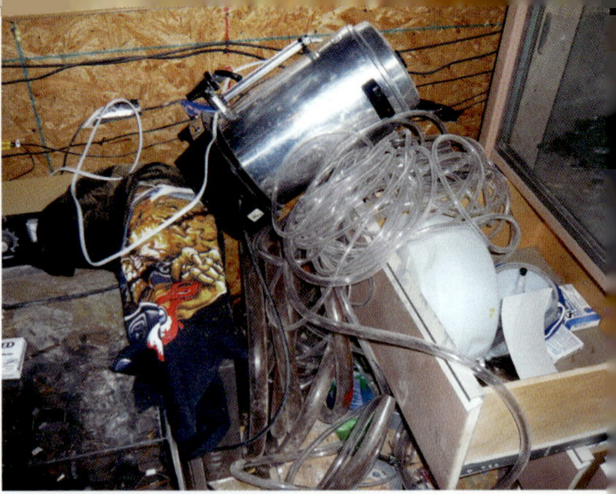
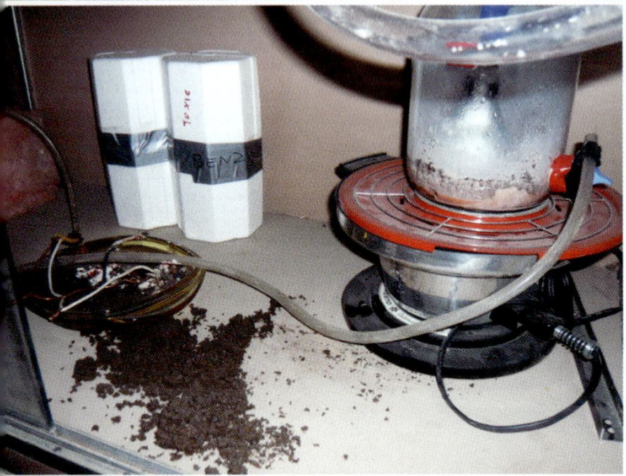
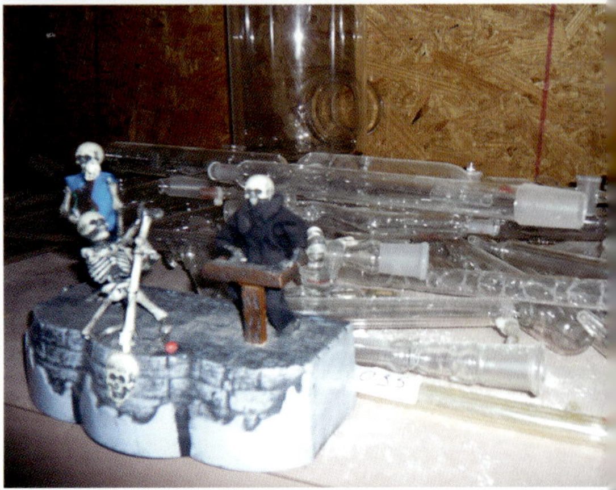
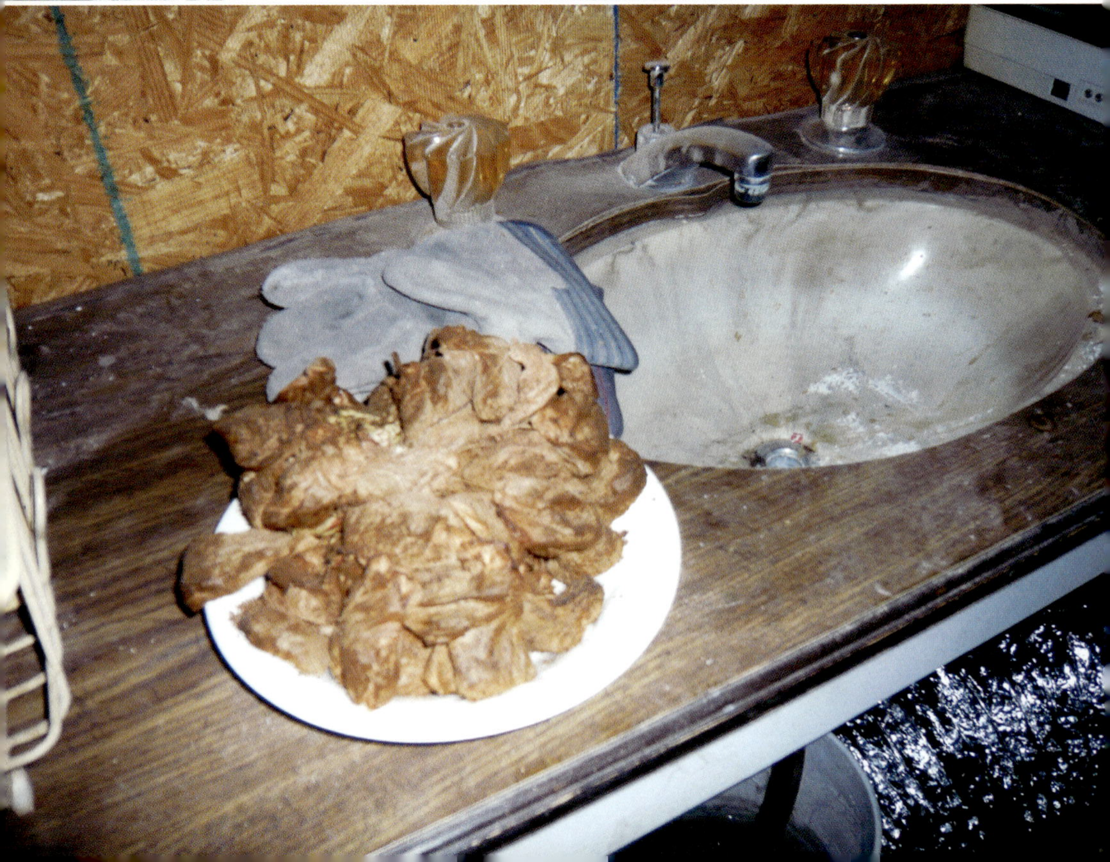

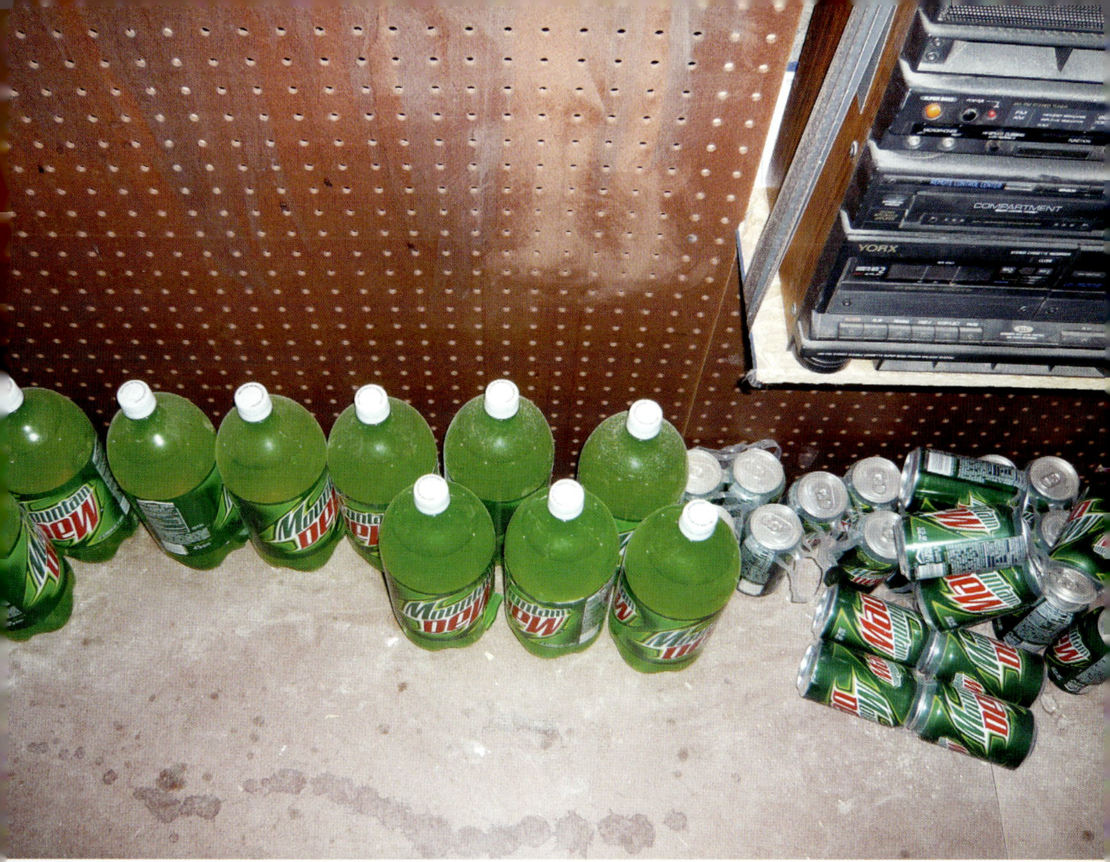
YORX
COMPARTMENT
Mountain Dew

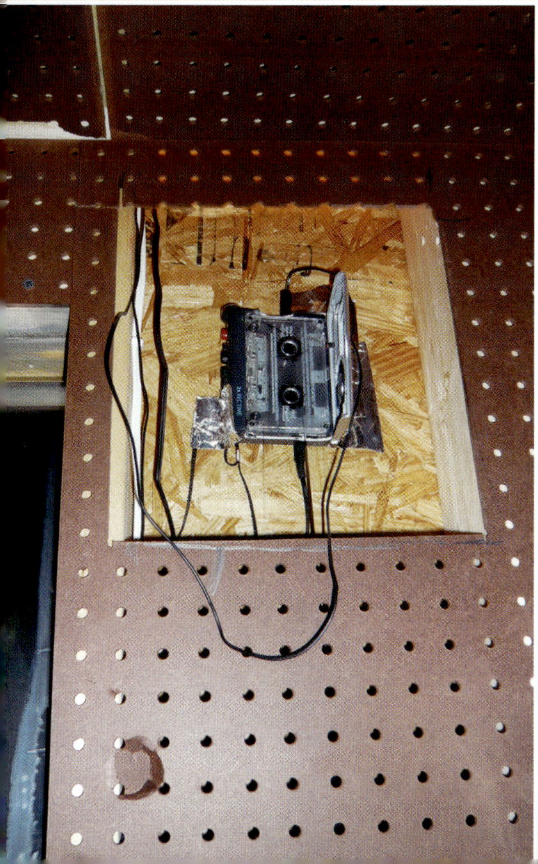

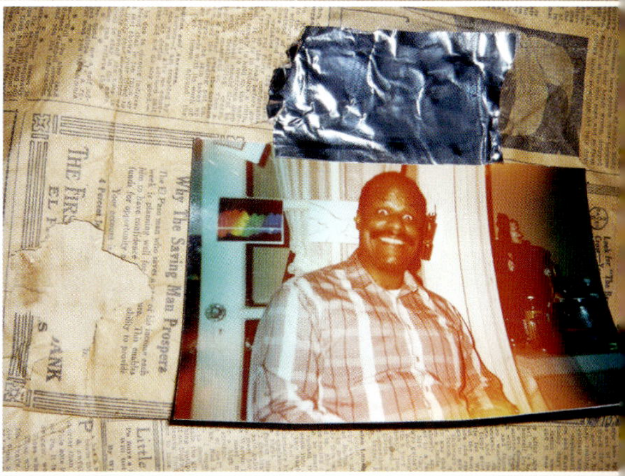
Why The Saving Man Prospers

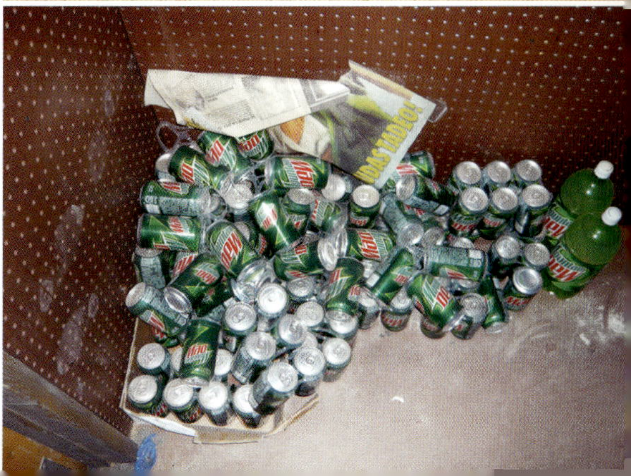

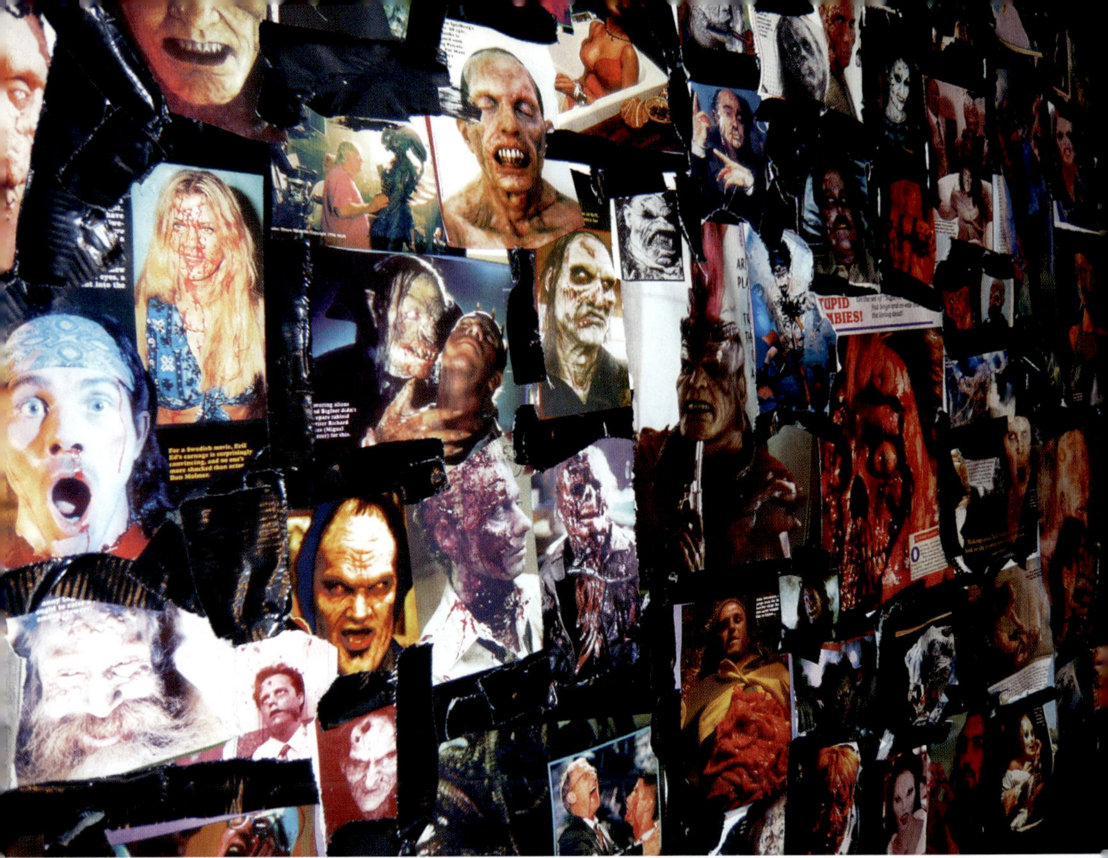

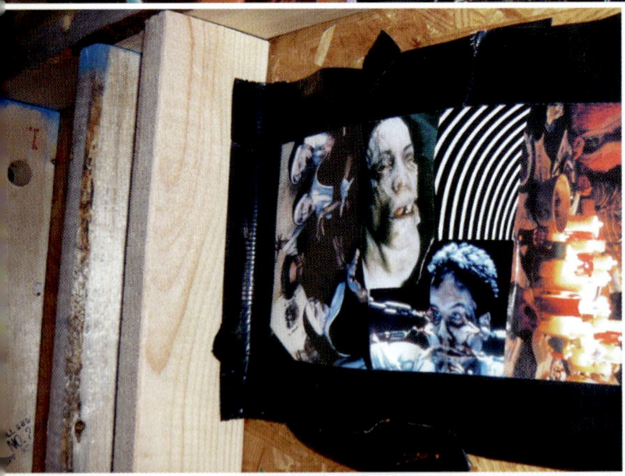

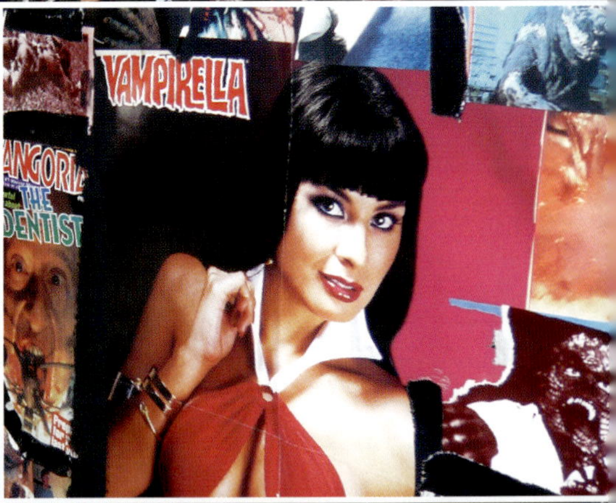
VAMPIRELLA

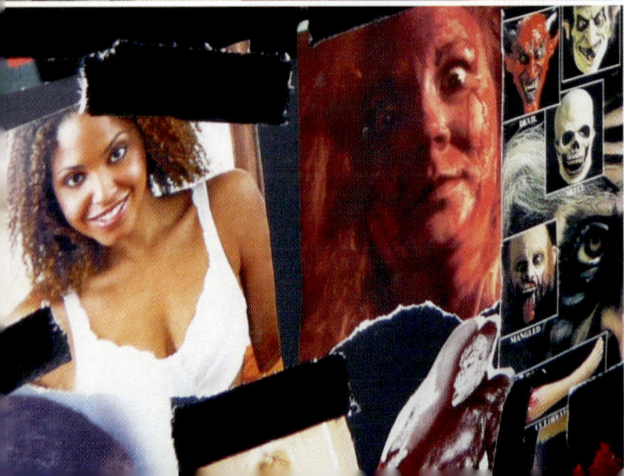

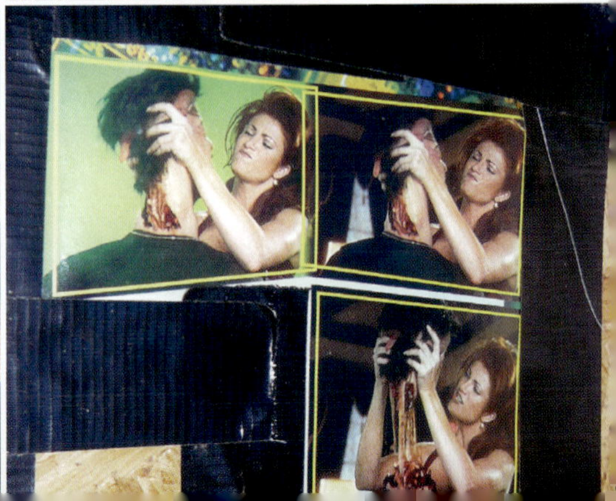

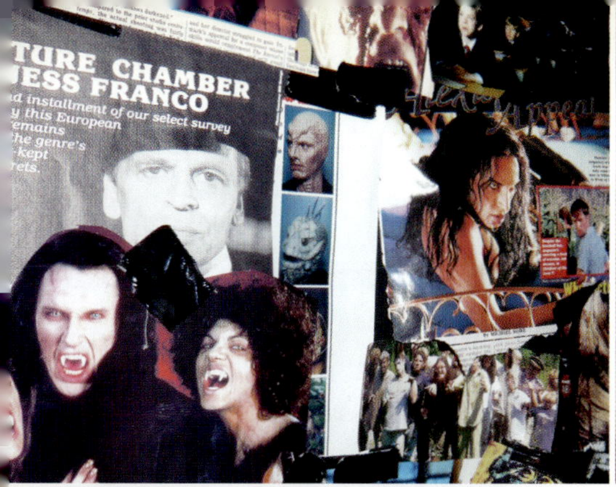
TURE CHAMBER
ESS FRANCO
d installment of our select survey
y this European
remains
he genre's
-kept
rets.

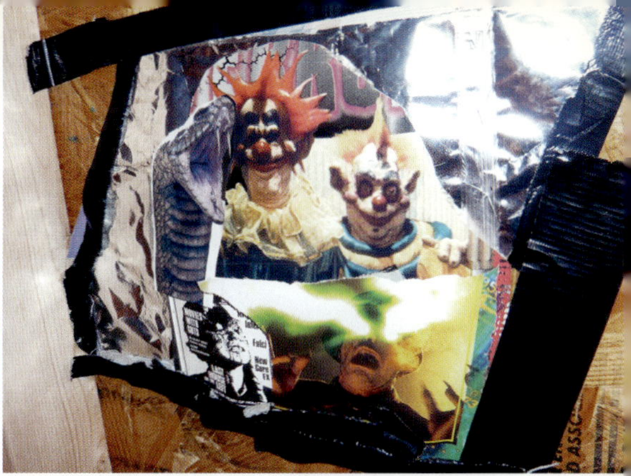

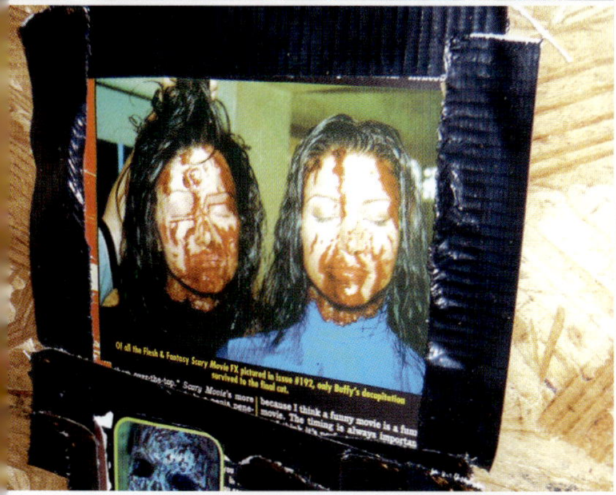
Of all the Flesh & Fantasy Scary Movie FX pictured in issue #192, only Buffy's decapitation survived to the final cut.

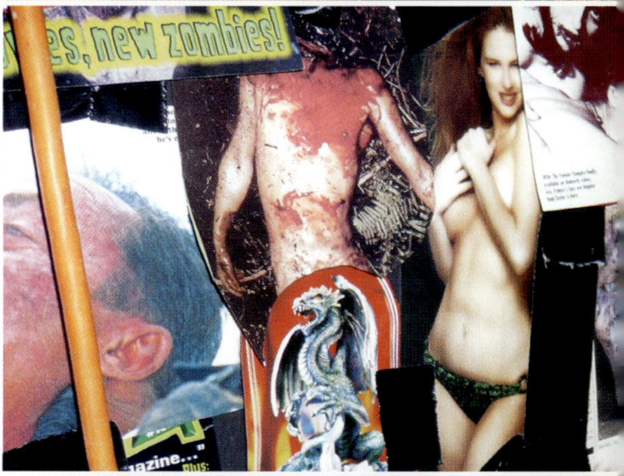
es, new zombies!

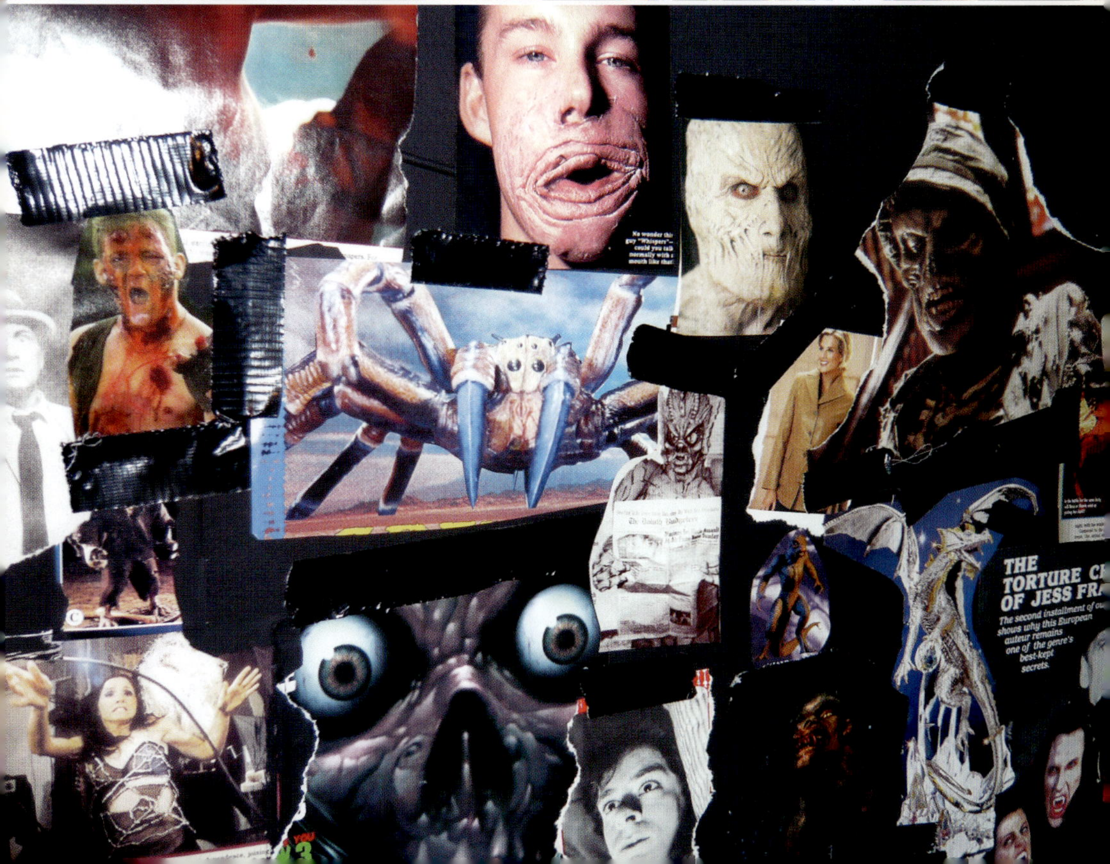
THE
TORTURE C
OF JESS FR
The second installment of ou
shows why this European
auteur remains
one of the genre's
best-kept
secrets.

Hello Meth Lab in the Sun

Jonah Freeman · Justin Lowe · Alexandre Singh

INSTALLATION PHOTOGRAPHY BY

Bill Diodato

BALLROOM MARFA

A publication in conjunction with the exhibition:
Hello Meth Lab in the Sun
April 5 – August 3, 2008
Ballroom Marfa, Marfa, Texas

BALLROOM MARFA
A NON-PROFIT CULTURAL SPACE

COORDINATED BY
Fairfax Dorn and Alicia Ritson

BOOK DESIGN & PRODUCTION
Paul Wagner

ISBN 978-0-9817586-2-6

PUBLISHED BY
Ballroom Marfa
P.O. Box 1661
108 East San Antonio Street
Marfa, TX 79843
www.ballroommarfa.org

AVAILABLE THROUGH
D.A.P./Distributed Art Publishers, Inc.
155 6th Avenue
New York, NY 10013
www.artbook.com

PRINTED IN HONG KONG

Contents

Introduction

To realize *Hello Meth Lab in the Sun* one had to cross the threshold of exhibition normalcy. An immersive installation, it consumed all who entered, commanding more of everyone than a passive gaze. These dormant environments bore evidence of past occupation, with memories of their past lives traceable beyond the overwhelming material mass of the installation, through jarring smells of a burnt-out kitchen and the putrefaction within the communal pantry. Sensorially pervasive, *Hello Meth Lab* was unforgettable, leaving all who visited in a state of bewilderment and disorientation.

Jonah Freeman, Justin Lowe, and Alexandre Singh took everyone involved on an epic journey to create this most extraordinary and ambitious installation. This catalogue serves as evidence of a truly singular project. It uncovers the thought processes channeled through the project's physical manifestation: a complex array of connections between contemporary society, human destruction, and industry, forged in part by generationally specific visions of utopia.

The artists' colossal treasure hunt to find the perfect retro wallpaper, motel door, mason jar, *Fangoria* magazines, chemistry glassware, fire-engine-red carpeting, railroad ties, dead refrigerator, generic drop ceiling tiles, industrial light tubes, random but particular family photographs, crystals, newspaper ads, *Terminator* movie poster, soiled mattress, sad furniture, typical dirt, prickly pear cactus, kitty litter, and so much more, began over a year before the artists began their two-month residency in Marfa. More than just mere props, each item acted as a crucial component in the altered experience that the exhibition offered, an experience that in turn became an intense reality of performative sculpture.

Ballroom Marfa was very fortunate to work with such imaginative, fearless artists. Only they had the power to dream up such amazing work. Their vision almost became a religion for all who helped with the project. Sleepless nights were endured not only by the artists but also by Project Manager Alexander Mills, Director of Operations Vicente Celis, Exhibition Coordinator Alicia Ritson, Gallery Manager Jonathan Mergele, Development Director JD DiFabbio, and the ever-present Lalo Baeza. A cast of dedicated volunteers were also integral to the success of the project, notably Erika Blumenfeld and Marie Ely, along with Kyle Knoedell, the artists' intern and studio assistant. The Brooklyn-based psychedelic rock group Psychic Ills marked the opening of *Hello Meth Lab in the Sun* with a performance at Liberty Hall—I'd like to recognize and give sincere thanks to Vance Knowles, Ballroom's music and film director; our expert lighting engineer, Rob Crowley; and dedicated sound engineer, Gregory Smelley, for their efforts in producing the show.

Countless others joined in the effort to make this project happen: architect Mike O'Meara donated his time to digitally render the building plan; Scott Williams and his crew worked earnestly after hours to create the foundation for the labyrinthine spaces; and the McDannald Ranch generously contributed local cacti, one of the exhibition's recurring motifs. As we prepared for the future life of the piece, Sally Coleman and Caitlin Murray meticulously catalogued every detail of the exhibition's contents as if they were undertaking an anthropological study of this transformed environment. And Tim Crowley, the City of Marfa, and the Thunderbird Hotel came to our aid with gracious understanding and patience, providing sites for temporary storage.

I want to also recognize Paul Wagner, the catalogue's talented designer, for staying with the project and keeping us on track to ensure we have enduring evidence of this renowned and historical exhibition; one which I hope will inspire generations to follow. The catalogue's brilliant contributors, Liam Gillick, Alison de Lima Greene, David Hollander, and Raimundas Malašauskas, have given form to the various physical, visceral, and cerebral aspects of this multifaceted exhibition. Their incisive and creative responses make this more than a didactic representation, creating another dimension for understanding and experiencing the installation. These texts benefitted from the artists' talks and interviews, transcribed by Mara Levy with skill and unfaltering patience. Bill Diodato, the exhibition photographer, must also be thanked for his commitment to the project, and for his consistently high standards while he documented an extremely difficult environment. Ballroom Marfa's own photographer, Tom Jenkins, also contributed in the early stages to the records of *Hello Meth Lab in the Sun* prior to his unexpected passing: his kindness and professionalism will be sorely missed.

Adam Helms, my beloved friend, deserves a special acknowledgment for his part in this project—it was he who initially introduced me to Jonah Freeman, Justin Lowe, and Alexandre Singh.

Finally, my sincere gratitude goes out to our devoted Board of Trustees; the Andy Warhol Foundation for the Visual Arts; the Brown Foundation, Inc., Houston, Texas; Peter Norton Family Foundation; and Ballroom Marfa members. We could not have realized this incredible exhibition without their commitment and financial support. Further, Ballroom Marfa and this exhibition would not exist without the core support and direction of Virginia Lebermann, President and Co-Founder, whose dedication to exploring and understanding culture is felt both within Texas and beyond.

Fairfax Dorn, *Executive Director & Co-Founder*, Ballroom Marfa

1

360° in the Sun

Alison de Lima Greene

We were somewhere around Barstow on the edge of the desert when the drugs began to take hold.[1]

You pull off the highway. Getting out of the car you might see a bicycle or two, a couple of parked cars hugging the shade. No one is on the street. To the right is an entrance papered over with fading advertising circulars from El Paso and a MasterCard/Visa decal. The narrow door lets you into a claustrophobic, dim hallway with a small receptionist's alcove. Faded astrological charts and snapshots are arrayed across the wall opposite, like a yearbook of long-absent residents. The hallway continues, taking you past travel posters, a dusty mirror, a closet, until you reach the first room, an abandoned domestic space, littered with rags, books, albums, magazines, an upended mattress, and a dead television set. Eye-popping wallpaper, red carpeting (stained), and a candy-colored canvas showing the chemical structure of methamphetamine indicate past attempts to decorate the space, while a discreetly placed microphone points to a more recent intrusion. Around the corner you will be confronted with further evidence of surveillance, as additional microphones, recording devices, and wires eat into the walls.

These are the first stages visitors encounter in *Hello Meth Lab in the Sun*, a complex, densely layered installation created by Jonah Freeman, Justin Lowe, and Alexandre Singh for Ballroom Marfa in the summer of 2008. Two years earlier the artists had proposed "a sequence of seemingly disparate environments, interpreted through the practice of alchemy. Each of the various rooms... will bleed and morph into each other. The totality of the installation will create a ricochet of ideas, a focus on the metamorphosis of materials: vegetables into broth, match heads into meth. An industrial, culinary, and narcotic revolution."[2] Detailing the scheme further, they referenced museum culture, specifically natural history dioramas and period rooms, as they laid out the framework for the three dominant components: an underground meth lab, a corporate trophy room, and an organic hippie cult kitchen.[3] These components continued to anchor the project as it was realized within the walls of Ballroom Marfa, but other elements were added as well, at times in an improvisatory spirit, as *Hello Meth Lab in the Sun* evolved into a more discursive project, existing in both an insistent present and filled with echoes of our immediate past.

Everything seems to be ready. Are you Ready? Ready?[4]

Like a filmic jump cut, the next room abruptly changes the tenor of your passage. The kitchen you're now standing in has been the site of a corrosive blast, and most likely you would like to leave this room as quickly as possible. Harsh shafts of daylight spotlight the blistered and blackened refrigerator, stove, and sink, appliances that seem capable of reigniting. There is a gothic humor here too. Reaching out of the oven door is a single chicken's foot, clawing its way back to air.

At this point the first component of *Hello Meth Lab* coalesces, and the ensuing rooms conflate the ordinary with the bizarre. Immediately following the kitchen is a brightly lit showroom displaying an array of exploded mannequin heads, evoking Ralph Steadman's adrenaline-fed illustrations to Hunter S. Thompson's *Fear and Loathing in Las Vegas*: "What sells, today, is whatever Fucks You Up—whatever short-circuits your brain and grounds it out for the longest possible time."[5] Off to one side is a bag of kitty litter (which can be used to smother the acrid smell of meth production), and kitty litter also spills from the burst heads. Past a vitrine filled with cacti, the next passage leads to the heart of the meth lab: stacked magazines create a bunker-like cell, while boxes upon boxes of over-the-counter medications—Sudafed, Advil, and Tylenol—share a table with a sprawling homemade still.[6] You can climb up to a sleeping loft/love nest, where the walls are papered with pages ripped from *Fangoria* magazines, and watch a yoga exercise video enacted by a figure in a gorilla mask.

As the artist Alex Hubbard has demonstrated, a virtual taxonomy of meth labs and the lives they encase can be culled from the internet. Parsing one such image, Hubbard suggests: "What is also important about this lab is the photo of the lake above the worktable. It recalls long sleepless nights, stressful mornings of cooking meth, listening closely for the sound of brakes, or the police, or perhaps a disgruntled partner to break down the door and shoot the manufacturer or burn them with chemicals. In spite of all that, this photo can serve as a sanctuary, a place to imagine as one waits for the binders in your pseudoephedrine to break down."[7] Freeman, Lowe, and Singh elaborate on Hubbard's reading, offering clues to their absent protagonists through a scattering of magazines, action figures, and photographs of muscle cars, as well as in the retreat offered by the sleeping loft. "The habitual users became 'freaks' lost in an obsessive personal mania," they observe. "In some sense the mentality of a meth addict could be seen as a mirror to that of a modern corporation: each on a singular pursuit at the expense of everything else."[8]

> *They came, but they resented it. They were full of excuses for coming, but there were coins in their eyes and sand in their mouths.*[9]

A door ushers you into a white room. Your degree of comfort depends on your familiarity with Upper East Side art galleries and VIP lounges. The carpeting is blood red and pristine. Wainscoting and paneling give measure and rhythm to the walls. You see a series of tastefully framed black-and-white photographs. Isn't that Gena Rowlands? Nico? And over there, that man, wasn't he in that movie with Mia Farrow? Vitrines encase monolithic architectural fragments, one of which is charred and warped as if salvaged from a burning building. Haven't you just seen this? A second series of photographs, now in color, further reprises your previous passage. Here the paraphernalia of a meth lab is captured through the flattering lens of a magazine stylist; in some instances pure anarchy seems to have invaded the lab as absurdly constructed experiments erupt in cascades of foam and boiling liquids. In others, each plastic bucket, canister, and tube assumes the status of sacred instrument.

Originally conceived as a trophy room, this segment of *Hello Meth Lab* departs significantly from the artists' 2006 project description. In place of the array of products associated with corporate drug culture Freeman, Lowe, and Singh planned to present, the artists introduce instead an insistently temporal shift. Where the first rooms of *Hello Meth Lab* loosely evoke the American underground of the past three decades, the dramatically staged black-and-white photographs specifically resurrect New York's café society, somewhere in time between Capote's Black and White Ball (1966) and Norman Mailer's fiftieth birthday bash (1973). Frozen in the flash of a paparazzo's bulb, partygoers either loll before the camera or mask their faces with cacti and crystals, as if engaged in some arcane ritual. The enshrined mural fragments elsewhere in the gallery suggests a further extension of this ritual, adding yet another layer of temporal complexity to the environment.

Can I stay here for a while?[10]

Stepping through a fissure in the wall you leave New York City for Drop City (c. 1969). Shelves of mason jars catalogue both food and artifacts: lentils, tinfoil, magazines, light bulbs, crystals. A sturdy ladder leads you to a large kitchen under a geodesic dome. You are first greeted by the sight of a coyote and a fox comfortably nesting together. Collages, assemblages, crystals, and dream catchers are scattered throughout, and you encounter yet again a series of astrological charts, restating the possible future of what has past. You missed supper, however: only dusty traces of food, a few cigarette butts, and an assortment of gourds remain. In the center of the table a reel-to-reel player unspools a continuous loop of tape—another winds a labyrinthine path around both levels—while two microphones await your response

The artists' specific evocation of Drop City, a Colorado commune that was founded in 1965 and largely abandoned by 1970, once again locates *Hello Meth Lab* in time and place. Created in an attempt to realize Buckminster Fuller's social ideals and the collective spontaneity of Allan Kaprow's Happenings, Drop City later became a symbol of failed promise in the writings of Ray Mungo and T. Coraghessan Boyle. For Freeman, Lowe, and Singh the environment is both inviting and suspect: "The ideas of freedom, openness and connection to the land were very attractive... But utopianism of any kind is still an ideology that is in negation of something else."[11]

The buttons said on and off, forward and backward. I caught on to that fairly fast. I don't think I'm so stupid as to erase what's on a tape.[12]

Exiting the Drop City rooms through a refrigerator door you find yourself in an undecorated space, empty but for two speakers wired to play back the sounds of your progress through *Hello Meth Lab in the Sun.* You were the object of surveillance inside this maze of corridors and tableaux. You were the protagonist. And it is this situational encounter between viewer and environment, history and memory, that brings your visit to a close.[13]

Pomegranates from St. Dominic's tree, Indian idols; a copper knife, with which Jews circumcise their children, together with the stone that belongs to it; one deer antler, which sweated blood in a synagogue during Lent.[14]

Hello Meth Lab in the Sun is not unique in its immersive detail, nor in its engagement with the viewer. Prior examples can be cited: Claes Oldenburg's garrulous storefront installations or Ed and Nancy Kienholz's arrested tableaux. More recently Ilya and Emilia Kabakov, Mike Nelson, and Christoph Büchel have created similarly elaborate site-specific environments that reenact the entropy of abandoned spaces, and the cinematic theatricality of these projects also informs *Hello Meth Lab in the Sun*.

What distinguishes *Hello Meth Lab*, however, is its central metaphor of alchemical transformation. Looking back to this pre-modern era of scientific and spiritual inquiry, Freeman, Lowe, and Singh uncover the generative wonder inherent in creative action, and their installation finds its most powerful precedent in the *Wunderkammern* of the Renaissance and Baroque eras. Created by princes, scholars, and merchants, these assemblages brought together *naturalia* and *artificialia*, natural wonders and marvelous artifacts, gathered from around the world. The eclectic and even mystical intent of these displays did not survive the Enlightenment, and the museum practices that evolved from *Wunderkammern* reflected instead the encyclopaedist's penchant for classification, a transition that has been described as a passage from wonder to knowledge.[15]

Freeman, Lowe, and Singh reverse this course. Each detail of *Hello Meth Lab in the Sun* is grounded in research and made concrete through physical labor. Early inventory notes for *Hello Meth Lab* spell out the artists' intent to riffle through both materials and concepts: "Potato, Raven's Fake Blood, Dusty Hot Plate, Scatter Philosophy, Fluorescent Mirrors, Cactus, Horror in Wallpaper, Glass Covered Sports, Reel to Reel with Waist High Plants."[16] The temporal fugue that both disrupts and unites each segment of the environment successfully mines both the mainstream and counterculture of the late 1960s and early 1970s, perhaps not coincidentally the era when Freeman and Lowe were born. Yet like a dream play in which objects and events recur and mutate, *Hello Meth Lab* is also insubstantial and fantastic, vivid in recollection and vital in presence. The artists have insisted that their project is not a political one: it is not about meth labs, drugs, exploitation, the failed visions of 1960s idealism, nor is it about surveillance, although all these themes run through the installation. These are but props and signifiers for a greater revolution proposed by Freeman, Lowe, and Singh, one that reconnects the mind-blowing nature of wonder to everyday experience.

Deepest thanks are due to Jonah Freeman and Justin Lowe for their ready answers to my questions about the evolution of *Hello Meth Lab in the Sun*. I would also like to acknowledge Fairfax Dorn, Alicia Ritson, and Ralph McKay in friendship and gratitude; their enthusiasm for this project was vital to the development of this essay.

Notes

1. Hunter S. Thompson, *Fear and Loathing in Las Vegas: A Savage Journey to the Heart of the American Dream* (New York: Vintage Press, 1971), 3.
2. Jonah Freeman, Justin Lowe, and Alexandre Singh, unpublished project proposal for *Hello Meth Lab in the Sun*, 2006. Courtesy of Ballroom Marfa.
3. Neil Young's 1969 "Cowgirl in the Sand" lyrics not only provided inspiration for the title *Hello Meth Lab in the Sun*, but each verse also found its echo in the tripartite installation.

4. Thompson, *Fear and Loathing in Las Vegas*, 198.
5. Ibid., 202.
6. Given the FDA restrictions now placed on selling these products, the medication boxes were meticulously fabricated by the artists and their assistants.
7. Alex Hubbard's "A Taxonomy of Meth Labs," was first delivered as a lecture at Whitney at Altria in March 2006. A version of this lecture was published in *Plazma* 29 (2007): 30–31. I am indebted to Charles Froelick for bringing this to my attention.
8. Freeman, Lowe, and Singh, proposal for *Hello Meth Lab in the Sun.*
9. John Leonard, "Happy Birthday, Norman Mailer!," *The New York Times*, February 18, 1973. See http://www.nytimes.com/packages/html/books/mailer-birthday.pdf
10. Neil Young, "Cowgirl in the Sand," 1969.
11. Freeman, Lowe, and Singh, proposal for *Hello Meth Lab in the Sun.*
12. Rose Mary Woods, on the 18½-minute gap in the Nixon White House tapes, November 8, 1973. *Time*, December 10, 1973.
13. Within a few weeks of the opening of *Hello Meth Lab in the Sun*, the playback function ceased to work. Nonetheless, the insistent presence of tapes and monitors throughout *Hello Meth Lab* successfully conveyed the artists' intent to give the impression that "one is always watching and being watched." Indeed, by happenstance, the silence of the recording devices evokes one of the most notorious moments in recent surveillance history, the 18½-minute gap in the Nixon White House tapes. Revealed to the public in the summer of 1973, this ominous lacuna fulfilled Norman Mailer's prophecy that the time had come to monitor "how far paranoia is justified." Leonard, "Happy Birthday, Norman Mailer!"
14. A description of the *Wunderkammer* created by Archduke Ferdinand of Tyrol, from Philipp Hainhofer's 1628 travel diaries. Quoted and translated by Mark A. Meadow, "Quiccheberg and the Copius Object: Wenzel Jamnitzer's Silver Writing Box," *The Lure of the Object* (Williamstown, MA: Clark Studies in the Visual Arts, 2005), 40.
15. Celeste Olalquiaga, "Object Lesson/Transitional Object," *Cabinet Magazine* 20 (Winter 2006–2007): 8.
16. Freeman, Lowe, and Singh, proposal for *Hello Meth Lab in the Sun.*

2

Restaging Intensity

Liam Gillick

> *"In fact their amazement and excitement were enormous. They expected a tomb, but they saw the most incredible and endless galleries of art. All was installed in spaces more beautiful than those of Versailles, in magical palaces where all extremes of climate were unknown. Countless lamps, some incredibly bright, others soft, glowed constantly through the blue depths. It was like daylight that would never become night. Of course the sight was nothing like we know now; we need to work hard to imagine the psychological condition of our poor ancestors. They had become so used to discomfort and inconvenience on the surface of the earth that it is hard to imagine their enthusiasm at the moment that they were saved from death and relieved of all their worries at the same time."*[1]
>
> —Gabriel Tarde, *Underground Man*

French Philosopher Gabriel Tarde's only work of science fiction was published in 1904. Yet it is a very contemporary tale. Globalization is followed by ecological catastrophe, leading the last few inhabitants of a frozen earth to reject the surface and construct a new architecture with no exterior—underground. The subterranean for Tarde is a place where art and philosophy reign. The underground as a total alternative—a permanently private zone of activity unrestrained by the relentless presence of surface nature. As such his book can be seen in the context of a culture that was accentuating art as an autonomous activity for its own sake.

His characters initially imitate nature, but as time passes they begin to produce a new art that is "better than nature."

A link exists between drugs, class, and industrial production in their accentuation or denial of "nature." Within spaces constructed for the production of synthetic drugs, there is a necessity to exclude the view. It is important to neither see or been seen from the outside. A new world is created, both in the site of production and in the head. Class systems control nature, apply their own naturalizing processes to any understanding of the way things are and, more importantly, the way things should always be. Industrial production requires control and a degree of secrecy. The recipe for homemade crystal meth is easier to find on the internet than the recipe for the production of Coca-Cola. Each of these categories carries its own aesthetic and its own veils against the cyclical relentlessness of nature. The cycles of nature are replaced by cycles of intoxication, production, reinforcement, striving, and perverse growth. It is these semi-autonomous concepts the artists in this exhibition deploy—embedding an artistic sensibility in an overwhelming series of structures that accentuate and exceed the natural.

The notion of a hidden space that reveals both inner and outer potential is at the heart of narcotic fulfillment. A meth lab is an excess of appropriation. Materials, time, knowledge, and the redesignation of elements. The drug of the title always implies a degree of crazed focus and delusion that it might actually be possible to make your own complex drugs. Speed gives you the energy to focus on complex tasks, up to a point. Meth is a self-generating concept. Taking it makes you want to stay up and do something. What better than deploying the excess focus in the production of more speed? In other words, the consumption of what is produced provides exactly the kind of jagged stimulus necessary to enter into the process of producing more. The producer and the consumer, often the same person, are locked in a cycle of excessive focus and production. A closed world emerges, of secrecy, delusion, and transformation. A perfect analogue of the artist's studio.

And a meth lab really is a lab. It is not a metaphor. The further appropriation of the laboratory as a nominative base for an exhibition challenges the notion of exhibition-as-laboratory that has been perpetuated as an experimental paradigm over the last fifteen years or so. A meth lab is a post-discursive space. It is a site of excess, silence, and danger. Using a meth lab as a model for an exhibition is more complicated than organizing a series of lectures or being contingent and imprecise about the potential of the art object. It carries less clarity in a standard critical sense. It explains less while showing more. The meth lab is always barely under control. It is a site where the aesthetic is provided by necessity. It is this aspect of the meth lab that has been unraveled here. A new methodology toward the production of a shattered art.

Despite the title, this is not merely a simulation of a meth lab. Of course there is a trashed kitchen, the signifier of the best party and the site of domestic production and rebuilding. A site to rebuild a motorbike engine, start a company, or write this text. The kitchen is always the logical site for the homegrown laboratory. But this kitchen is also a site for the generation of ideas and a critical challenge to formalist and super-subjective art. This kitchen functions as a moral locus within the analysis of desire, coding, and projection. It is also the first place to look when you raid a house.

This exhibition is a carefully executed thing but that does not imply that it is the result of the careful study and reproduction of a series of real sites or actual situations. In this case the artists are not showing us what such spaces of production, flake out, and delusion look like; they have taken on the frame of mind that creates the incidental aesthetic of intoxication. With the possibility that there was some parallel form of intoxication taking place at the same time.

The role of the viewer in this case is strained. We are not being shown a tableau vivant. Nor a true diorama in the museological sense. In fact, the exhibition as a whole has more in common with the perceptual shift familiar from science fiction. A dose of Tarkovsky or even Kubrick is cut with the techniques of repetition and clarity following chaos that we know already from the battle between angst and irony at the heart of much contemporary art.

This exhibition offers a number of spaces each of which reveals more of the artists' complex reading of process and desire, while at the same time stripping away the viewer's set of reliable reference points. We should come to accept this work not as a challenge to any values in particular, but as a series of corrupted and reimagined spaces, each of which exceeds—in terms of symbolic signification—and simultaneously avoids operating as a mirror of that which we already know.

Hello Meth Lab in the Sun does not revel in the display or re-creation of excess for its own sake. Instead this exhibition searches for new markers and sites of production and attempts to offer them as zones for the rethinking of art, in particular and in general. All elements could be stripped out and presented separately, but there is an important point here about collective practice. We do not imagine ourselves in a series of spaces beyond art, but rather in a series of new places that have been redefined by artists. Maybe this process has taken place without checking, and that's the key. A documentary described down the phone and produced with no degree of verification or care. If this is the case then the same is true for the art component of its aesthetic DNA. A lack of documentary rigor is matched by a collapse of artistic logic; these twin attacks are the essence of the project. The viewer or potential user of the work is locked in a perpetual interplay between the semiotic textures of the work "as art" and the misleading evidence of devastated sites of class, excess, decay, and delusion.

Notes

1. From Liam Gillick's artist's book *Underground Man (Fragments of Future Histories)* (Brussels & Dijon: Michéle Didier and Les presses du réel, 2004), updated from Gabriel Tarde's original of the same title, first published in English in 1905 by Duckworth, London.

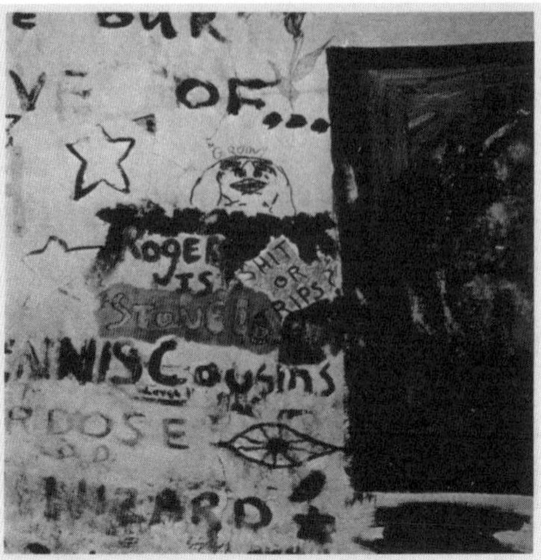

Graffiti on a dining room wall of a ghetto house has accumulated over months and is the product of the efforts of a series of occupants. (Photo by Nancy Sterngold)

Fig. 1

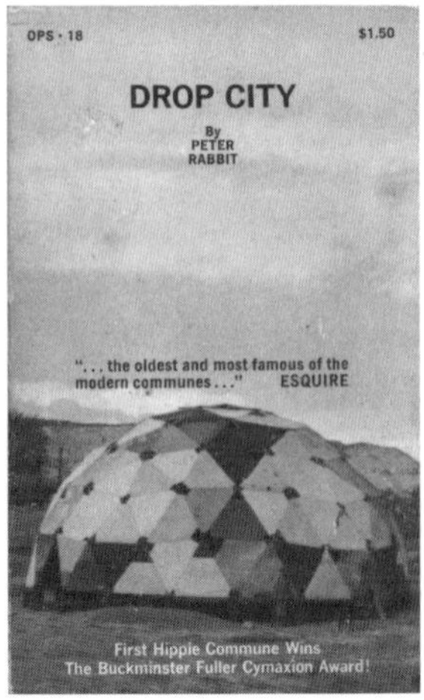

Fig. 2

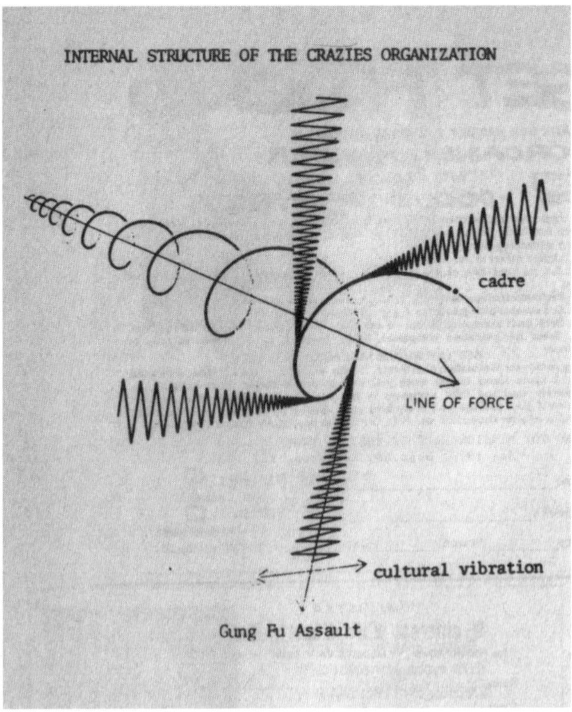

Fig. 3

3

AN 3219: Community Ritual and Group Psychosis

David Hollander

In this six-week course, we will explore some of the many ways culture is derived from biology, and the stabilizing and standardizing influence it has on human social relations.

Architecture without Architects
What is "Architecture"? Is "Architecture" still a useful concept? Or rather, could "Architecture" ever be useful? What kinds of things might replace "Architecture"?

Ethnopharmacologic Search
Just as the loss of inhibitions and the achievement of a shared psychic state are felt to be the products of a rap session, insight is felt to be achieved by ritualized drug use. The world is quite different when one is high.[1] [Fig. 1]

The Effigized Landscape
Magic is made manifest through the dilapidation of the symbolic order. A two-way radio set made entirely out of bamboo will function identically to one made in a factory in Asia, but additionally be capable of realizing extra-dimensional contact.

Hexing
Hexing is a conceptual game that every human being is playing every time he opens his mouth; but very few people are aware of the game. This class will provide you with the history, technique, structure, and applications of hexing. The extent to which one can control his hexing is the extent to which one can control his universe.[2]

Wolf Children and Feral Man
This class will explore Transition and Transformation while living in communal antinomian anticipation of a new world order.

After We Die, What Then?
An examination is made of the phenomenon of Ska Maria Pastora and the Re-emergence of the Fertility Cult in the New World. A discussion of Plant World Science Fiction will follow. SF reading list is attached.

SPECIMEN TEXTS
A. *Drop City*, by Peter Rabbit [Fig. 2]
B. International Werewolf Conspiracy Manifesto
C. *Mescal, and Mechanisms of Hallucinations*, by Heinrich Klüver
D. Diagram: Internal Structure of the Crazies Organization [Fig. 3]

EXAMINATION
Sunday—Bible Class, usually starts at 1:00 PM

Reading passage specimen A
I invented a spaceship. It was a box with a hole in the middle. I got in. I had a stick with an old clodhopper shoe glued on one end. The shoe was stuffed with rags and newspaper to make it just like a foot. I invented this spaceship.

There was a hole in the bottom. I took the stick and stomped the ground with the shoe as hard as I could. It was too hard. The spaceship blasted off and was going too fast. I couldn't control it. I was going up into the sky really fast, too fast. There was a huge blue wall in front of me. It was the sky. I couldn't slow down, I was going to crash into this solid blue wall, the sky. It was everywhere. But then I just crashed through. Made a hole in the blue sky the exact same size as the box. And then everything was black and full of stars. I was in outer space, headed for the moon.[3]

Reading passage specimen B
We are the ultimate Horror Show...Hideous Hair and Dangerous Drugs... Armed Love striking terror in the vacant hearts of the plastic Mother and pig-faced Father...

A small town filled with comfortable merchants and their well-fed wives lies twinkling in the valley as the clouds drift peacefully across the night sky. In the distance a wolf howls, and another one answers, and another, and soon the night is filled with the roar of howling wolves. The sounds merge into a chaotic chorus as the wolves gather under the fullness of the moon... Lights go on all over the town. Gentlemen in pajamas rush to their windows, unable to sleep because of the horrifying sound carried by the night wind. Women in curlers clutch the blankets close around themselves, as the shadowy, howling shapes move through the streets.

When morning comes, there is silence. The wolves have gone away, have moved on in their mysterious way... Their stomachs sag with the weight of a satisfying feast and their fangs drip with the fresh blood of their victims. THE CHILDREN OF THE TOWN RUN BESIDE THEM.

One of the wolves stops long enough to say, "Survival."

And another one answers him, "REVOLUTION."

The worst fear is the fear of the unknown, and we are the Unknown...[4]

Reading passage specimen C

The barking of a dog moved the whole picture and vibrated through my right foot. This was so distinct I thought it necessary to identify the dog with my right foot.[5]

Directed Writing and Composition Examination

The following questions having been created to test your comprehension of the preceding specimen texts. In formulating your responses, keep the preceding specimen texts in mind, and refer to them specifically whenever possible.

Answer **all** questions.
The number of marks is given in brackets [ ] at the end of each question.
Dictionaries are **not** permitted.

Explain what you understand about the term "Mainstream American Life." [2]

In what ways is a dolphin like a computer? [4]

In what ways is a cactus like a book? [4]

Explain what you understand about the term "Alchemy." [5]

In your own words, briefly explain how Lycanthropy works. [2]

SF READING LIST

- John Wyndham. *The Day of the Triffids*. New York: Doubleday & Company, 1951.
- Brian Aldiss. *Hothouse*. New York: Simon & Schuster, 1962.
- J. G. Ballard. *The Drowned World*. New York: Berkley Publishing Corp., 1962.
- Frank Herbert. *The Green Brain*. New York: Ace Books, 1966.
- Thomas M. Disch. *The Genocides*. London: Ronald Whiting & Wheaton, 1967.
- John Boyd. *The Pollinators of Eden*. New York: Dell Publishing Co., 1969.
- Ursula K. LeGuin. *The Word for World is Forest*. New York: Berkley Publishing Corp., 1972.
- Alan Dean Foster. *Midworld*. New York: Ballantine Books, 1975.

They crawled away from their great and beautiful towns, left their cities, deserted their houses—as if all at once home had suddenly become alien. Their social structure also collapsed, and all organization died overnight. From then on, the weeds began to flourish in the streets, and the pollen to blow over the cash register; the advance of the jungle had begun.[6]

Notes

1. William L. Partridge, *The Hippie Ghetto: The Natural History of a Subculture* (New York: Holt, Rinehart and Winston, 1973), 50.
2. David Felton, *Mindfuckers: A Source Book on the Rise of Acid Fascism in America* (San Francisco: Straight Arrow Books, 1972), 120.
3. Peter Rabbit, *Drop City* (New York: Olympia Press, 1971), 109.
4. *BAMN (By Any Means Necessary): Outlaw Manifestos and Ephemera, 1965-1970*. Stansill, Peter and Mairowitz, David, eds. (London: Penguin Books, 1971), 167.
5. Heinrich Klüver, *Mescal, and Mechanisms of Hallucinations* (Chicago: University of Chicago Press, 1969), 50.
6. Brian W. Aldiss, *Hothouse* (London: Faber and Faber, 1962), 142.

4

In Conversation with the Artists

Raimundas Malašauskas

Raimundas Malašauskas: So who are those people with cactuses, crystals, and coyotes in the black-and-white photos?

Jonah Freeman: The way we described it initially to the people that we were collaborating with for the photo shoot was that it was a cross between Norman Mailer's fiftieth birthday party and the last scene in *Rosemary's Baby.* We wanted to have a sense of a kind of affluent urban crowd that gave the aura of wealth and power, but not exactly a corporate situation, something that was a little more along the lines of Radical Chic... like the people at Norman Mailer's birthday as captured by Garry Winogrand—that was a strong point of reference for the style of the pictures.

Justin Lowe: They might be some kind of board, like a Black Acid Co-op. We were thinking that these people were the ones that would be indicted on more of a global scale. Like they were the global chemists that might actually benefit from the production of Sudafed and its illicit use in making crystal meth.

Alexandre Singh: The pseudo-museum in which we displayed the photos had a certain ambiguity. It was possibly a museum "looking out," a space built by a culture to display its own artifacts and history. In that sense the photos also carry this same ambiguity; we don't know if they are being taken by these characters for their own pleasure or by an outside agency documenting this ritualistic activity. The title of the series—*Age Grade/Age Set*—is taken from the anthropological term for the stratification of a society along lines of seniority, particularly with reference to initiation rituals.

Raimundas: They look more like the shareholders of the company that produces Sudafed, rather than just individual drug designers or nonchalant users.

Jonah: Yeah, exactly. And then so as far as the cacti and crystals go, they were meant to imply some sort of ritualistic activity—hence the *Rosemary's Baby* reference. During this uptown cocktail party there is perhaps some abstract ceremony happening. Perhaps it is a sacrament or something of that nature. The cacti and crystals were intended to be very open signifiers. The implication here is that these materials have mystical powers. Crystals of course are believed to have the ability to transform the energy around them. And I think something like 80 percent of cacti have some sort of psychotropic effect on human beings.

Justin: So what we have is a mock anthropological depiction of a group of sophisticates being treated as if they were a tribe ingesting ethnogens in hopes of achieving a unified vision, a collective consciousness in order to empower and structure their tribe. For this particular tribe, however, these materials simply need to be combined in a certain set and setting in order to produce these transformative effects. And then there are linguistic associations like crystal meth or the crystal power, crystal technology, crystal healing, crystal balls—you get the drift.

Raimundas: Or *The Crystal World* by J. G. Ballard. And those people seem very calm and uptight. They don't seem to be in any way in a psychotic state. They are more involved in repeating a certain ritual as a way of being together. As a sort of community tool. They seem to me to be outside of psychotic loop.

Jonah: Yeah. We did not want to imply that these people were overtly psychotic at all. Perhaps they control the psychosis.

Justin: Or benefit from the psychosis.

Alexandre: Of course, from the outside it may seem weird, but to those involved in the activity it quickly becomes completely normal. The madman never thinks that he is mad.

Raimundas: In the photographs inside the installation I see that an accident happened there, some kind of explosion. It immediately makes me think of the conflation of things, when substances, forms, and processes get melted into each other. Even in terms of the conversation about it, rather than analyzing the rooms of the installation as separate entities representing specific groups I would rather see it as a mash of spaces and attributes.

Justin: That is on the mark because in this installation, although there are around fourteen rooms, there are three particular anchors, which are meth culture; the aristocratic society or industrialist society, which is partly manifested in the affluent group of people that we were just speaking about as global chemists or shareholders that have the means to produce a preservation society; and then there are the hippies who ran to the hills in hopes of inventing a new society.

Jonah: I do think a kind of mash-up occurs between these environments associated with these seemingly disparate groups. By putting them in a sequence we try to point toward some sort of macro-connection between them. For instance the hippie countercultural revolution of the '60s definitely opened up a sort of permissiveness as far as drug use goes; this could somehow be linked to the methamphetamine epidemic in America. Although meth culture is not an idealistic mind-expanding culture, it is still a ritualistic drug culture as far as these groups of people are producing very powerful drugs and then consuming them in some kind of nihilistic ceremony. And of course both of these groups are linked to the various sites of industrial power and production. Or it could be said they are enveloped by them. An example in meth counterculture is the easily accessible industrial products that go into the production like Sudafed and muriatic acid. In hippie culture it is a very direct reaction against industrial civilization. In their attempt to disengage and separate one could say they shine a spotlight on industrial society. In *The Hippie Ghetto* by William Partridge he speaks about the hippie people being defined partly by an obsession with the repugnant conditions of twentieth century America.

Justin: And this of course results in intentional utopian communities like the Hog Farm or New Buffalo, but what was interesting for us, because of their link to art history, was Drop City, which was named that because of the performances that took place at the University of Kansas where this group would drop objects onto the sidewalk as people passed below them. We riffed on this by making a homegrown geodesic duplex, accessed through a hole in the Upper East Side museum wall, which is experienced as a sort of fever dream.

Alexandre: The constant thread through these different narrative realities was this engagement with alchemy. In the utopian communities' cooking, in the dystopian meth production, in the sculptural elements there is this constant theme of material transformation.

Raimundas: It is like the different sections exist simultaneously although they are coming from different periods. In that sense it reminds me of period rooms in the museum, when you have a room in the style of Napoleon II, and then Napoleon III, and then behind the next door you can find some Bauhaus furniture because this is how history is narrated in this museum. So in that sense the whole display of *Meth Lab* was based on museum logic.

Jonah: Yes, definitely. In the way that museums can offer a form of time travel. But they can also be like a graveyard. The idea of presence through absence is an important component for all the rooms in the piece. Spaces that were once inhabitated but are no longer.

Justin: This is an essential thought, and I think it really led to our material choices, not only in how they relate to each different community and their environments, but also the inherent aura or residue of material, for instance in voodoo practices how items from the people you want to contact or hex must be used in the ritual, and some things have a larger capacity for mojo than others.

Jonah: You could say that we were going for some sort of materialistic fiction. In the sense that the central theme of community, ritual, and psychosis is strictly manifested in the traces of activity on an environment.

Alexandre: It is interesting to consider how that museum metanarrative affects our reading of the environments. Obviously they are always simulations, and much more so than any cinema or theatrical set. Is the museum itself abandoned? Maybe these are parts of a museum that have been occupied by the same group the museum is cataloguing.

—

Raimundas: Did you have a list of all the substances that you were going to include in each room?

Alexandre: There was a process of writing lists, then compiling and amalgamating them. Then writing them out again. There was a line-budget which was a massive Excel spreadsheet. It covered everything from the construction of each room down to, say, five fake feathers and three dried gourds.

Justin: I mean, community ritual and psychosis could very well have easily applied to us when we were making this.

Raimundas: As well as the rites of passage of the art world, because I think it is not only meth, but also like a meta-lab, i.e., something that illuminates its own conditions of making and what is beyond.

Jonah: This idea of the meta-lab is interesting. I don't think we thought so specifically about how this would relate to the community of the art world. The one thing we did think about is how the theme of alchemy could be applied specifically to sculpture or the act of making sculpture. That sculpture, at least in a traditional sense, is using one material to represent another. Marble becoming flesh, for instance. This is not exactly lead into gold, but it is a transformation. The material of fantasy. Or back to the materialist fiction idea.

Justin: And while we weren't trying to use materials as stand-ins for something else, we were trying to combine different materials in hopes of arriving at something else. In some cases we did not always know where it was going to end up, and it was quite explosive. While we were aware of the products in the experiment, we were not always aware of the result. For instance, mustard gas being the unfortunate by-product of a chemistry project gone wrong.

Raimundas: It's also like the customization of mass industry. You customize a magazine or other mass-culture product: for example, mix Mentos with Pepsi-Cola, and make an explosion.

Jonah: Meth production is definitely a customization of mass culture. A real abject DIY mentality.

Justin: Well that is the funny thing about the experimental attitude of a particular drop-out, hippie intellectual subculture that was really into the Transcendentalists. You go from Thoreau's cabin to Kaczynski's cabin and then *blamo*, the sound of confusion.

Raimundas: That's a good point. Did you include any sound element in the project?

Alexandre: Yes. The whole space was wired for sound, with dozens of mics embedded in walls, in plant pots, in kitchen cupboards. In fact there were two *Terminator* posters, each with a mic sticking out of Arnold Schwarzenegger's eye.

Jonah: As you moved through the space these mics would be recording you. The recorded sound would pass through various analog tape loops built into the walls of the structure, either a retrofitted cassette Walkman or a reel-to-reel would create a playback delay. So the recorded sound would then be sent to another part of the installation usually three to six minutes later. Ostensibly you could say something in one room and hear yourself being played back in another. This leads into a sort of paranoid surveillance culture, but at the same time retains a DIY mentality. Like how a tweaker might do surveillance.

Alexandre: Of course the mics also recorded these distorted playbacks, which again were played back into the spaces three to six minutes later. After a while these washes of echoing corrupted sounds, hisses, clicks, and faint voices would bathe the space. Someone's passage through the installation would create artifacts that could drift around the various rooms for hours.

Justin: It was experienced as a locationless audio hallucination.

Raimundas: The sound was going in a loop, right?

Justin: It's a loop, but it's a degenerative loop, so each time it passed through it became fainter and fainter. And so the idea was that it would always continually record over something else. So it sort of was almost like the same kind of technology a dub studio might utilize.

Raimundas: Exactly. Or the technology you used to connect liquid containers with pipes, creating a looped feedback system in itself.

Alexandre: Exactly, the distillation still. Researching all the glassware that we used for the laboratory I learned more than I'll ever need about all the different kinds of ways to purify a substance. It is quite beautiful how, for example, a Cold Finger Condenser loops back on itself. It seems that there are a million different ways for the snake to eat its own tail.

Justin: The connection between different objects is sort of a continual theme, whether we're plugging chords into cacti, or we have tubes bringing in sunlight from the outside that are then kind of broken up to redirect the sun from different areas, or microphones that go into a speaker in another part of the building.

Raimundas: So there were lots of intersecting, overlaid loops in terms of space, sound, and time-experience.

Justin: Yes.

—

Raimundas: Do you have a long-term interest in alchemy?

Alexandre: For myself, yes, very much so. I'm interested in all aspects—from the original writings of Hermes Trismegistus to the hermetic tradition of Giordano Bruno to the caricatured alchemists of fiction, our various Fausts and Dr. Frankensteins. The later evolution of the tradition into theosophy, the Magic Circle, Aleister Crowley, and so on is where we link up as a group, especially with Justin's interest in 1960s counterculture and the West Coast contemporary interest in mysticism and cults.

Raimundas: To me, the interesting link of your work with alchemy can be found in the article by Arturo Schwarz on Marcel Duchamp and alchemy: he was seeing the whole idea of the Readymade basically coming from alchemy. For example, a pissoir of Duchamp ideally represents the idea of what a philosophical stone is, because it's about taking something that looks like anything else, the most mundane object and then finding transformative qualities in it. I think you treat industrial chemistry and mass-culture detritus in a similar way.

Alexandre: Remember also that the pissoir's primary function is to receive the golden stream of a human being. Later Duchamp has it transform itself into aesthetic gold. You may laugh but there was a long held belief in medieval times that urine did in fact contain gold and there were serious attempts undertaken to extract it. They were literally trying to 'take the piss'.

Jonah: I think the concept of alchemy as an underlying theme was born out of our discussions. This really gets into the nature of how the collaboration worked. We spent nearly eighteen months just bouncing ideas around that would then bring about lots of unforeseen connections between our individual interests.

Justin: There's also this idea of alchemists operating in a secretive underground way, in some ways like a contemporary counterculture. They never really shared their recipes with other people, but a lot of these people discovered basically the same things. And then if you go through and you look at the texts and diagrams there's an overwhelming amount of similarities and parallels. However it was never like there was some kind of alchemists' conference, you know. They weren't meeting in the Holiday Inn to discuss their findings for the year.

Alexandre: Ha! Yes, that's true. Those similarities originate from their shared source material. This material was written in a deliberately esoteric form, so ambiguous that anyone could read whatever the hell they wanted from it. You could just shake your head and say "Oh sir, you know nothing of the mighty art, this symbol actually represents something completely different." Some people might argue that the continued success of contemporary art resides in a similar system of interpretation. There are no real answers; there are only questions!

—

Raimundas: Is the building of a spectacular environment, this installation replete with the attributes of utopian communities and cultures, a genre exercise?

Justin: I don't think that it's a genre exercise. I think that we're too far along in our individual practices and this is just a combination of them, you know. I mean, Jonah's obviously very involved in the making of movies and fictions, and I'm very involved in the making of things that could operate as sets for movies. So we're kind of pulling together things in a very cinematic form. This isn't an exercise in a genre that we aren't already involved in. This isn't our country album or something.

Alexandre: Well, yes and no. I think we're all aware of the history of this kind of work. From Kurt Schwitters' *Merzbau* to Mike Nelson's *The Coral Reef*, there is a trajectory of exploration in environmental narratives. As an artist you're interested in exploring a convention and pushing it into new territories. I'm not sure that "country album" isn't such a bad label after all for this piece. Or perhaps this is our funk-country-speed-metal-fusion album?

Raimundas: Is this the first thing you've done together?

Jonah: Justin and I had been making collages together: but very casually. All three of us were sharing a studio, so there was naturally a lot of discussion about potential ideas for projects and a general engagement with each other's work to the extent that at some point it seemed that all three of our practices could come together in some sort of collaborative project. Then a friend of ours, who was aware that we had been throwing around these ideas in the studio, suggested that we make a proposal to Ballroom. So we coalesced a bunch of what we thought would seem like ridiculous ideas—things no one would go for, and we put that into a PDF proposal. We really took our past endeavors in environmental installation and pushed it somewhere strange.

Justin: We were living with each other and drew a lot of sculptural ideas from our surroundings, mainly Grand Street in Brooklyn, which is a bunch of different shops with very idiosyncratic storefronts that exist in a sequence—so it is quite cinematic—which resulted in a collective germination for the sequential structure of the rooms in *Hello Meth Lab in the Sun*.

Jonah: And the general structure of civic life in New York City could be seen as a kind of mash-up. You are constantly being confronted with unlikely architectural combinations and unlikely social groups stacked on top of each other.

Raimundas: I once had an experience in L.A. when I went to The Museum of Jurassic Technology, and I spent a couple of hours inside, in this whole strange, strange universe. And then the museum was closing, and I got outside to the street, and I felt like *whew*, I'm back to reality. But then I looked around and it was L.A. and I suddenly realized that I'm not in a reality that is more real than the inside of The Museum of Jurassic Technology. I'm in Los Angeles, and it feels like just another layer of The Museum of Jurassic Technology. It's just another floor. It's just another level. To me it was sort of like this glimpse of the essence of L.A. in a way, but I needed to enter and exit The Museum of Jurassic Technology to have that glimpse.

Justin: Well I guess that's sort of what you hope for with one of these kinds of pieces. That it does carry on, that you have a hangover from it, and that you see the world for a little while through the project's lens. And that it is something you can't ward off.

Jonah: I've had that experience with really good films. When you leave the theater the tone and world of the film is so well drawn that it bleeds into the non-cinematic reality.

Alexandre: I think that effect operates on people outside of art, films, and so on. I think many normal spaces like hospitals, schools, canteens, offices, etc. breed their own alternate realities. The social environment, the fluorescent lights, the lack of an outside perspective creates a psychosis on the people inside. We always wonder how someone could "go postal." Just visit a Brooklyn post office for a couple of hours—I think it's pretty obvious how. And then there's the counterpart of the guy in the log cabin out in the wilderness, the survivalist in his bunker. It's something that we played with in the implied narrative of spaces like the pantry and the orange wallpaper lounge.

Justin: That'll happen in a mall for me. And I know this is a real future cliché but… I saw that movie *Synecdoche, New York* and then I walked out and I was in a food court and then I turned to the left and there was a navy recruiting center and then I step outside and I'm in a college campus. And so it worked really well. I kind of liked that more than the film.

Biographies

Jonah Freeman was born in 1975 in Santa Fe, New Mexico, and lives and works in New York City. Recent solo exhibitions include: *Black Acid Co-op* (with Justin Lowe), Deitch Projects, New York (2009); *In the Kaleidoscope Room*, Mitterrand + Sanz, Zurich (2009); *Hello Meth Lab in the Sun* (with Justin Lowe and Alexandre Singh), Ballroom Marfa, Marfa, Texas (2008); *The Franklin Abraham*, Andrew Kreps Gallery, New York (2004); and *In the Public Realm: Sixteen Scenarios*, Public Art Fund, New York (2002). Recent group shows include: *Paper Exhibition*, Artists Space, New York (2009); *Hello Meth Lab with a View* (with Justin Lowe), The Station, Miami (2008); *The Future as Disruption*, The Kitchen, New York (2008); *Grow Your Own*, Palais de Tokyo, Paris (2007); *Day Labor*, PS1/MOMA, Long Island City, New York (2005); and *Vanishing Point*, Wexner Center for the Arts, Columbus, Ohio (2005).

Justin Lowe was born in 1976 in Dayton, Ohio, and lives and works in New York City. Select solo exhibitions include: *Black Acid Co-op* (with Jonah Freeman), Deitch Projects, New York (2009); *Hello Meth Lab with a View* (with Jonah Freeman), The Station, Miami (2008); *Freedom Time Is Here Little Kitten*, Galerie Frédéric Giroux, Paris (2008), Galleria Cesare Manzo, Rome (2007); *Hello Meth Lab in the Sun* (with Jonah Freeman and Alexandre Singh), Ballroom Marfa, Marfa, Texas (2008); *Helter Swelter*, Oliver Kamm/5BE Gallery, New York (2006); *In Practice*, Sculpture Center, Long Island City, New York (2005); PS1 Special Project Room, Long Island City, New York (2004). Selected group exhibitions include: *Fit to Print*, Gagosian Gallery, New York (2007); *Slow Burn*, Galerie Edward Mitterand, Geneva (2006); *The Pantagruel Syndrome*, Castello di Rivoli, Turin (2006); *Greater New York 2005*, PS1/MOMA, Long Island City (2005); and *The Melvins*, Anton Kern Gallery, New York (2003).

Alexandre Singh, was born in 1980, in Bordeaux, France, and lives and works in New York City. Solo exhibitions include: *Assembly Instructions*, Jack Hanley, San Francisco (2008); *UNCLEHEAD* (with Rita Sobral Campos), Museu EDP, Lisbon; *Hello Meth Lab in the Sun* (with Jonah Freeman and Justin Lowe), Ballroom Marfa, Marfa, Texas (2008); *The Marque of the Third Stripe*, White Columns, New York (2007). Selected group shows include: *Performa Biennial*, New York (2009); *A Certain Realism*, La Galerie, Noisy-le-Sec, Paris (2009); *Fax*, The Drawing Center, New York; *Of this tale…*, Royal College of Art, London (2008); *EAST International*, Norwich Gallery, Norwich (2007).

Contributors

Alison de Lima Greene is Curator of Contemporary Art and Special Projects at the Museum of Fine Arts, Houston. Among Ms. Greene's recent publications are *Isamu Noguchi: A Sculpture for Sculpture* (New Haven: Yale University Press, 2006) and "Ellsworth Kelly and the Large Wall," in *A Fine Regard: Essays in Honor of Kirk Varnedoe* (London: Ashgate Press, 2008).

Liam Gillick is a British artist working in New York and London. His artistic production is multidimensional. Through writings and the use of specific materials in his artworks, Gillick examines social, political, and economic systems. Gillick will represent the German pavilion at the 53rd Venice Biennale.

David Hollander is a filmmaker and collector who lives in Marfa, Texas. He produces artist-made films, most recently for Stephen Prina, Mathias Poledna, Erin Cosgrove, and Jennifer Lane.

Raimundas Malašauskas is a curator of contemporary art and a writer currently living in Paris. His works evolve around contemporary phenomena, persons, biographies, and stories, addressing the parallel worlds of science, media, film, literature, and mass culture.

Bill Diodato is a photographer who explores the fine line between art and commerce. Diodato's editorial work has been showcased in a wide range of publications including *Interview*, the *New York Times Magazine*, *Nylon*, *Vibe*, and *Vogue*, among many others. He has photographed campaigns for Clinique, Clairol, Cole Haan, Prada, Victoria's Secret, and Marlboro. Diodato is currently working on his first gallery exhibition.

Acknowledgments & Credits

THANKS TO
The Andy Warhol Foundation for the Visual Arts; The Brown Foundation, Inc., of Houston; Peter Norton Family Foundation; The Cowles Charitable Trust; Foundation for Contemporary Arts; and Ballroom Marfa members for their generous support for *Hello Meth Lab in the Sun*.

IN-KIND SUPPORT PROVIDED BY
Meghan Gerety & Michael Phelan
Billy Marginot
David McDannald
Michael O'Meara

THE ARTISTS' GALLERIES
Andrew Kreps, New York
Monitor Gallery, Rome, Italy

HELLO METH LAB IN THE SUN
PROJECT TEAM
Fairfax Dorn, *Director*
Alexander Mills, *Project Manager*
Alicia Ritson, *Exhibition Coordinator*
Jonathan Mergele, *Gallery Manager*
Vicente Celis, *Director of Operations*
JD DiFabbio, *Director of Development*
Sarah Burkhardt, Sally Coleman, Kyle Knoedell, Caitlin Murray, *Project Interns*

ADDITIONAL INSTALLATION TEAM
Scott Williams Construction
Naegele Construction
Steve Holzer

VOLUNTEERS
Erika Blumenfeld, Jon Coleman, Jeff Elrod, Marie Ely, Mary Farley, Isaac Hollander, Lauren Klotzman, Robin Lombaria, Karen Longshore, Nicole Marie Paterson, Dustin Peavey, Steve Rabourn, David Tompkins, Cory Van Dyke

NEW YORK CITY PHOTO SHOOT CREW
Rebecca Conroy, *Production and Casting*
Jessie Cohan, *Stylist*
Tim Daley, *Digital Tech*
Jeff Vallee, Kyle Knodell, Sarah Burkhardt, *Photo Assistants*
Sarah Smart, Sebastain Quinn, *Production Assistants*

ADDITIONAL THANKS TO
Mona Lisa Furniture and Hassan Salmeh

BALLROOM MARFA STAFF
Fairfax Dorn, *Executive Director*
Vance Knowles, *Director of Film and Music*
Alicia Ritson, *Associate Curator*
Jonathan Mergele, *Gallery Manager*
JD DiFabbio, *Director of Development*
Vicente Celis, *Director of Operations*
Sally Coleman, Caitlin Murray, Park Myers, *Ballroom Marfa Interns*

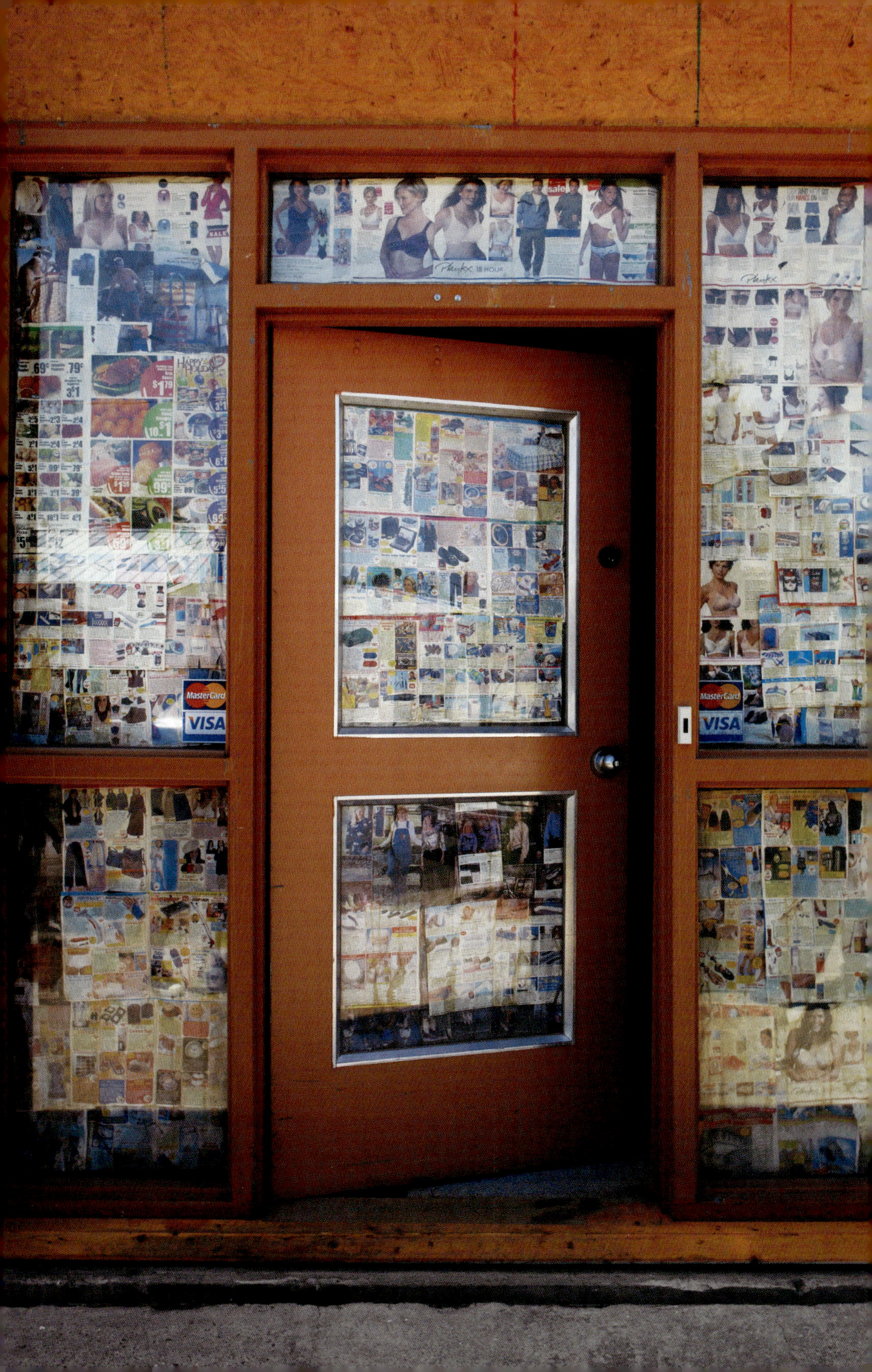
MasterCard
VISA
MasterCard
VISA

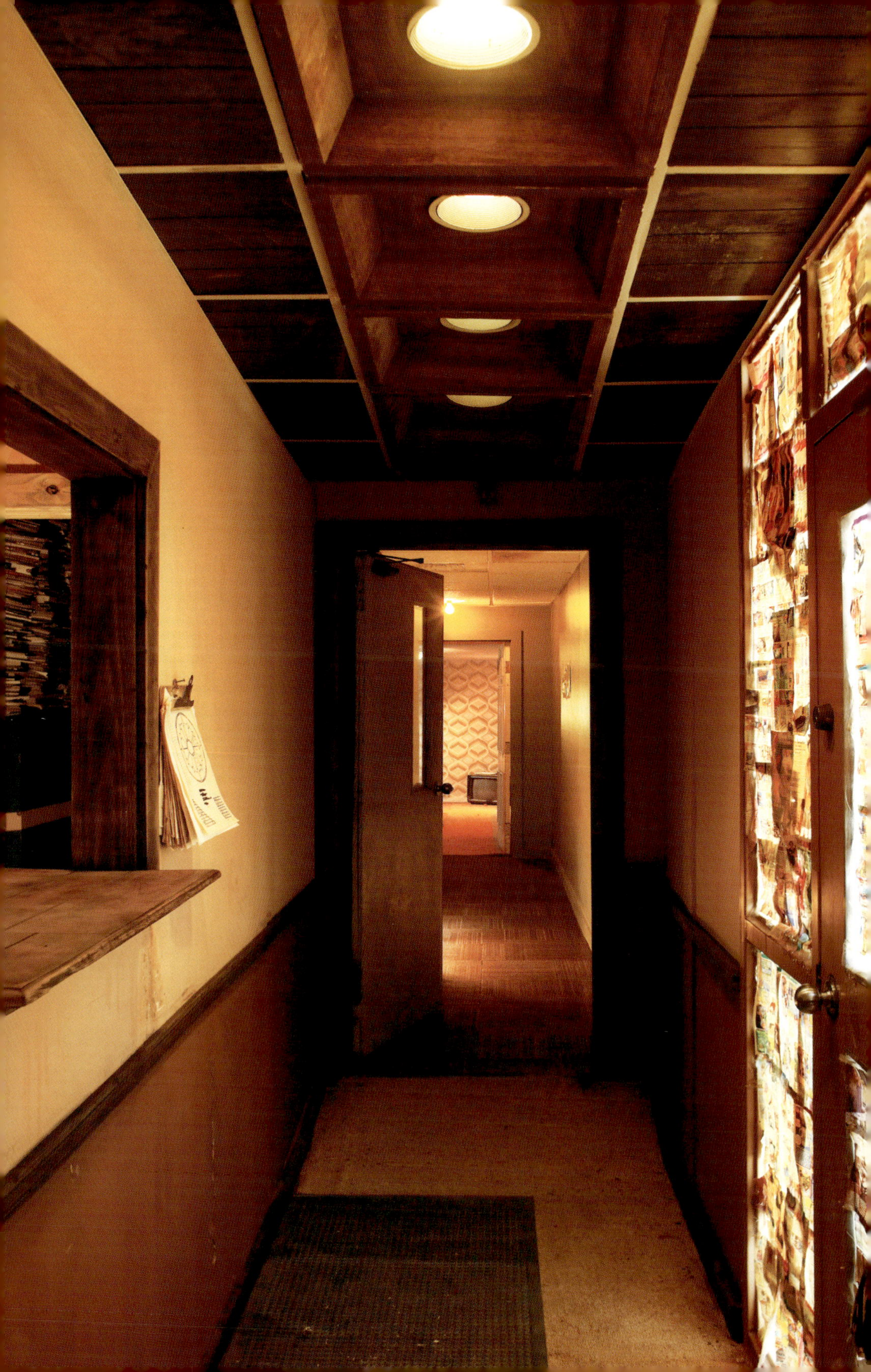

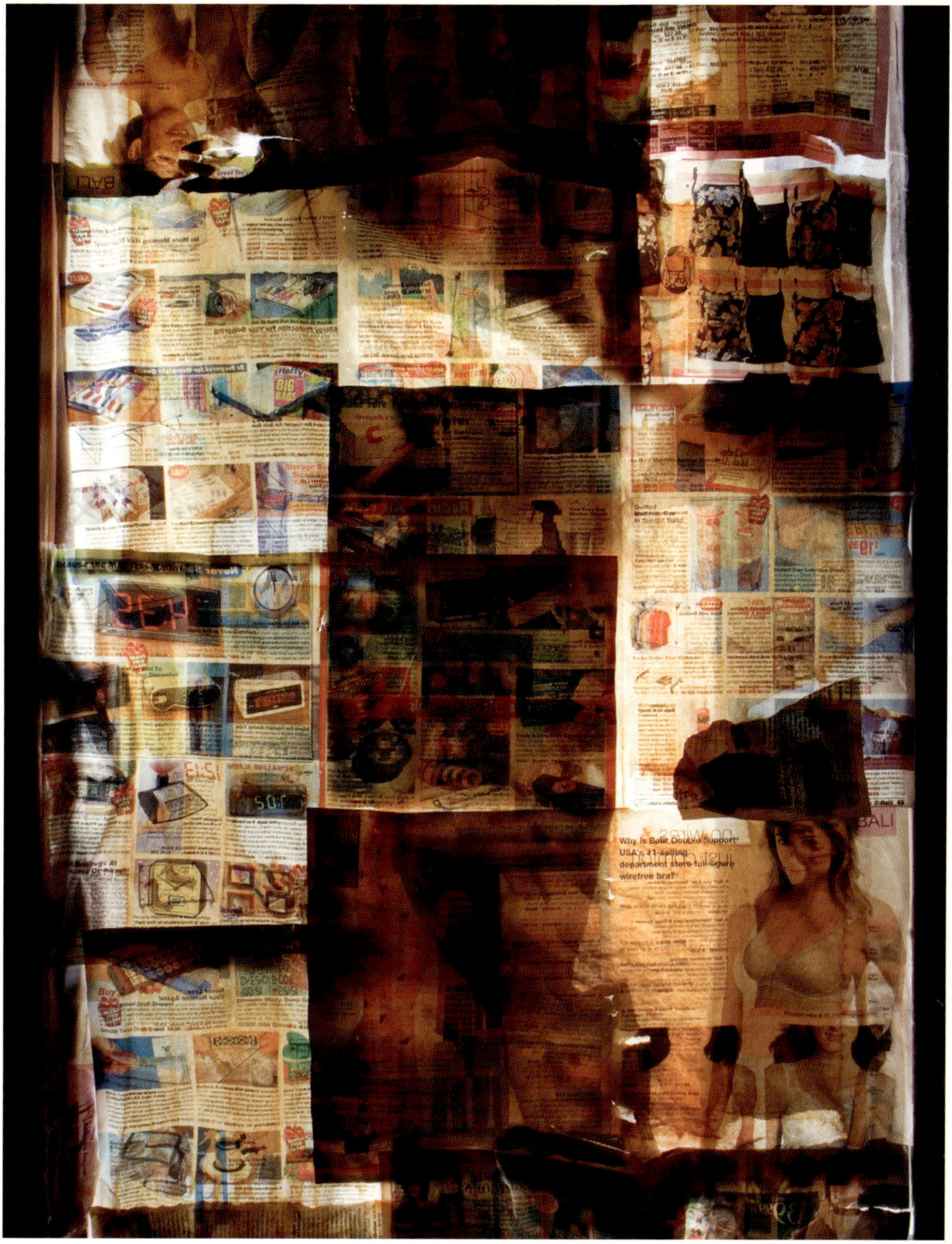
BALI
Why Is Bali Double Support® USA's #1-selling department store full-figure wirefree bra?

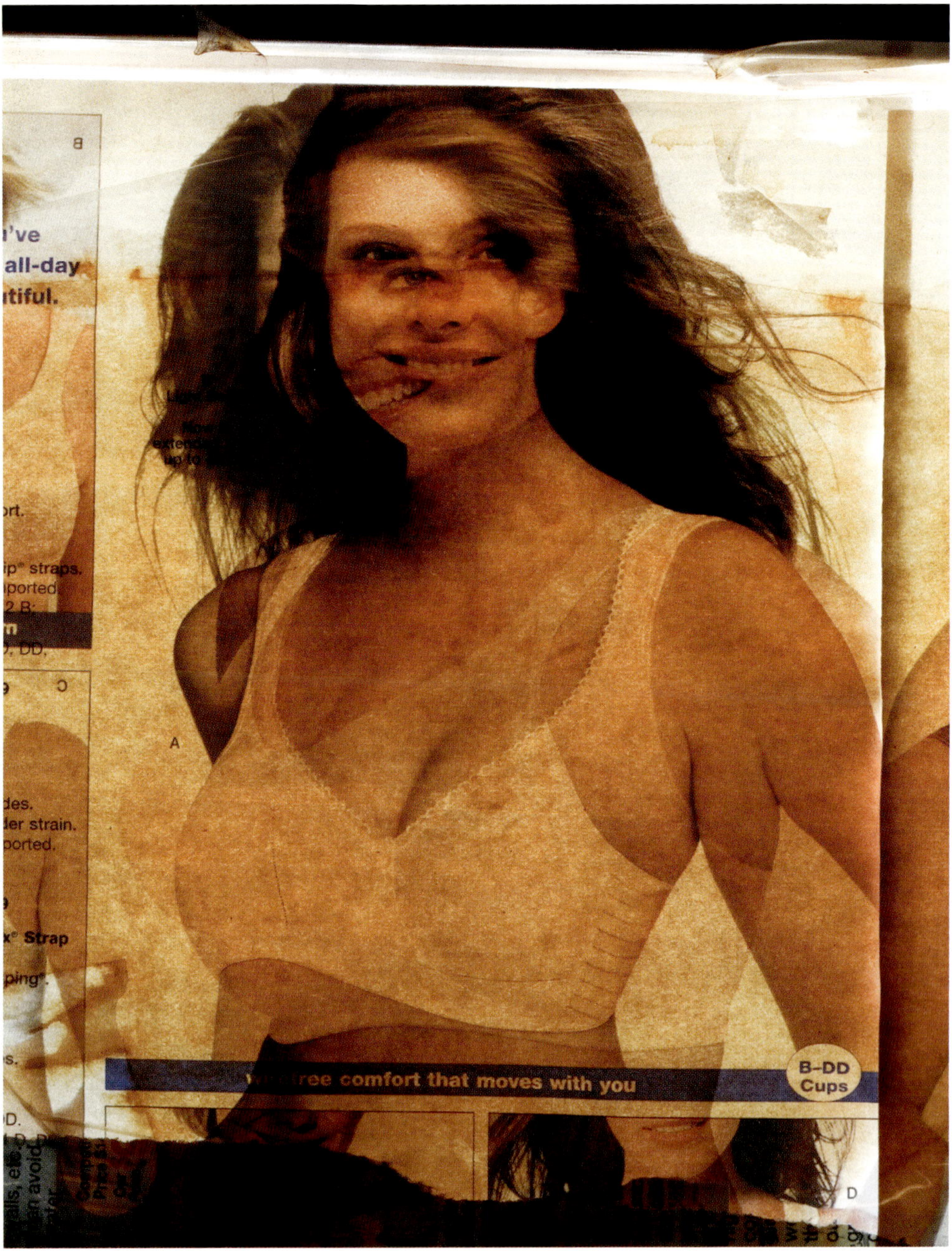
all-day
comfort that moves with you
B–DD
Cups

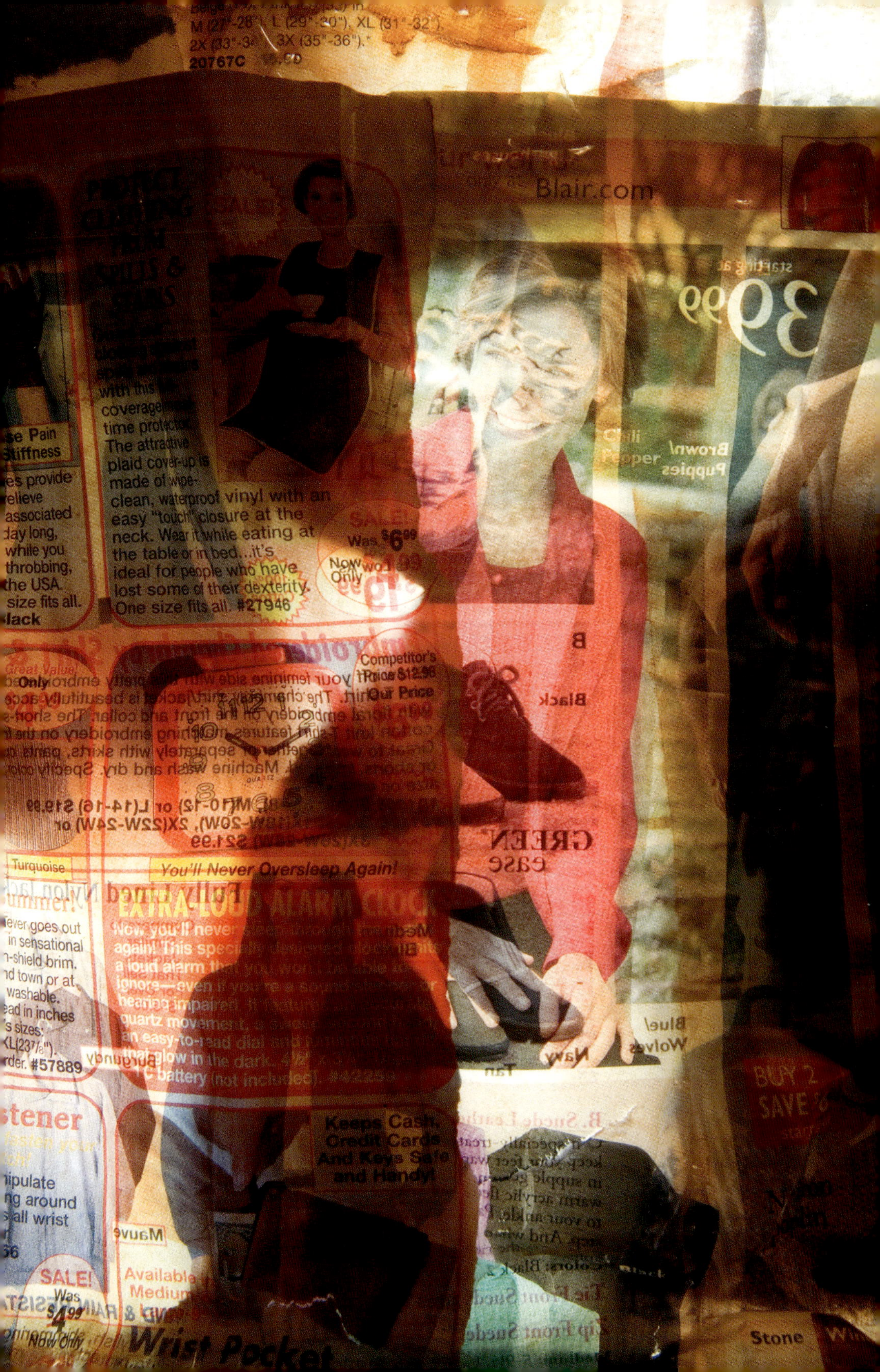

M (27"-28"), L (29"-30"), XL (31"-32"),
2X (33"-34"), 3X (35"-36")."
20767C
Blair.com
starting at
3999
Chili Pepper
Brown/ Puppies
with this
coverage
time protector.
The attractive
plaid cover-up is
made of wipe-
clean, waterproof vinyl with an
easy "touch" closure at the
neck. Wear it while eating at
the table or in bed...it's
ideal for people who have
lost some of their dexterity.
One size fits all. #27946
SALE!
Was $6.99
Now Only
Competitor's Price $12.98
Our Price
Black
You'll Never Oversleep Again!
EXTRA-LOUD ALARM CLOCK
Turquoise
Burgundy
Keeps Cash, Credit Cards And Keys Safe and Handy!
Mauve
Blue/ Wolves
Navy
Tan
BUY 2 SAVE
Stone
Wrist Pocket

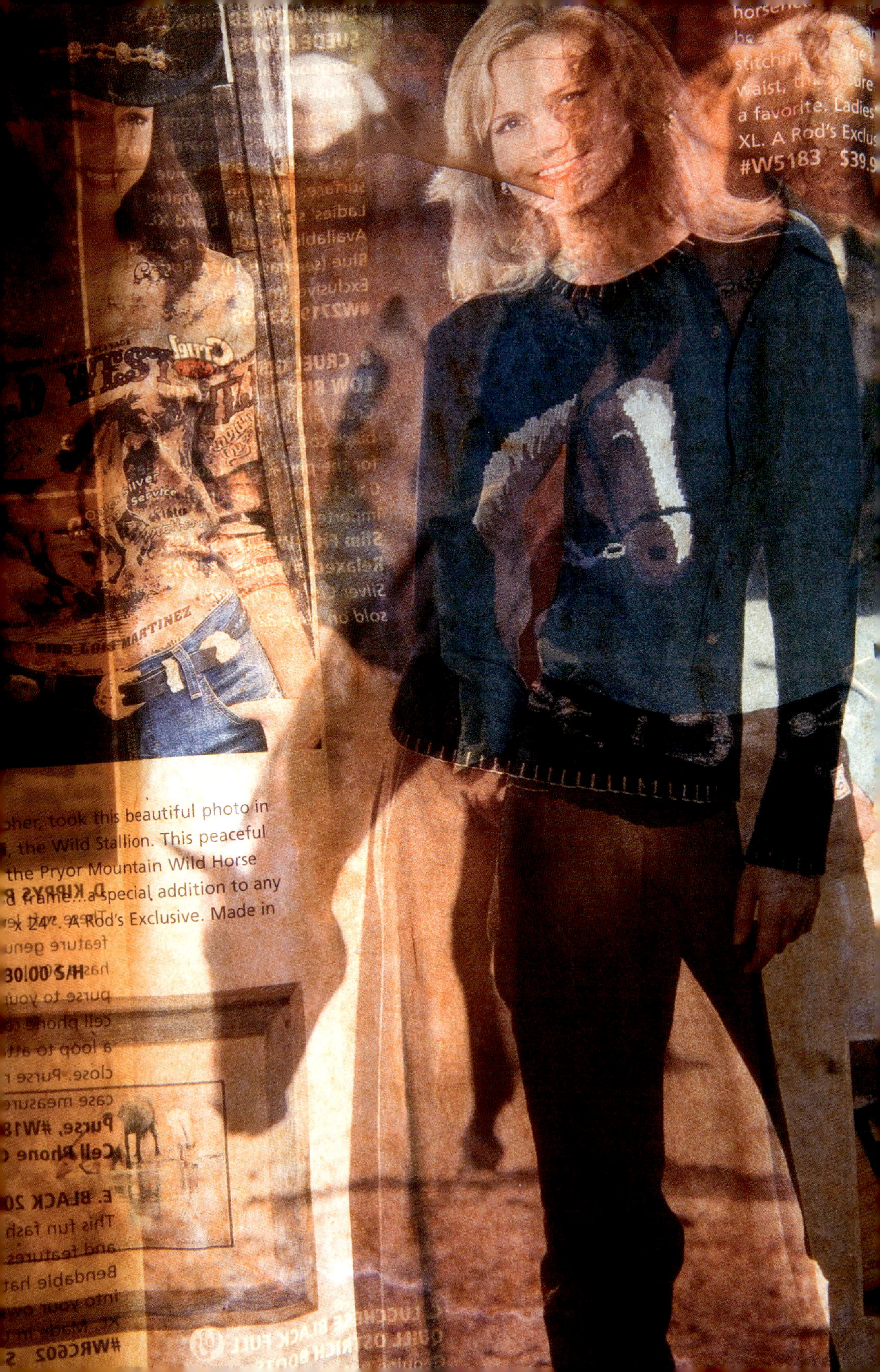
stitching
waist,
a favorite. Ladies'
XL. A Rod's Exclus
#W5183
took this beautiful photo in
the Wild Stallion. This peaceful
the Pryor Mountain Wild Horse
frame...a special addition to any
x 24". A Rod's Exclusive. Made in

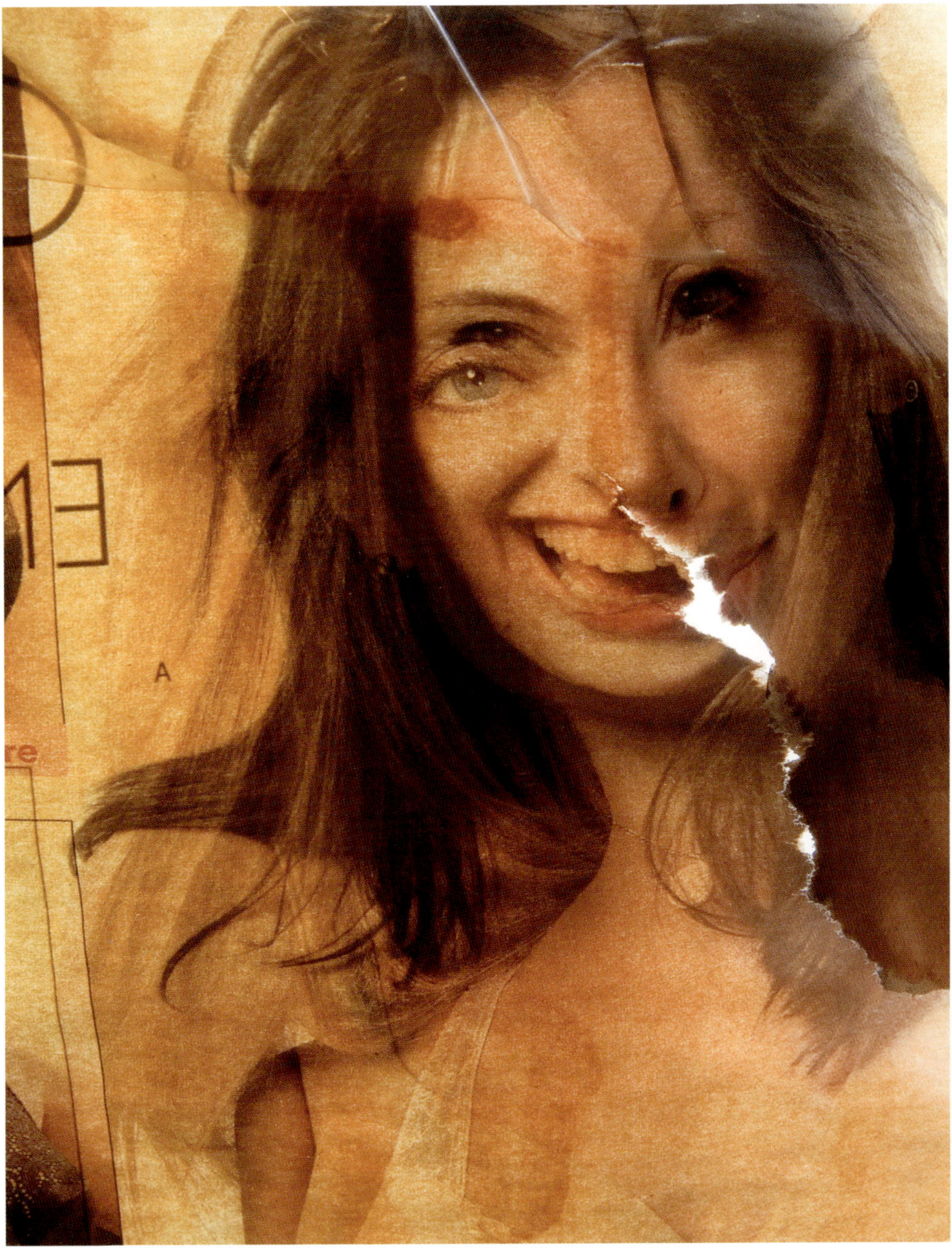

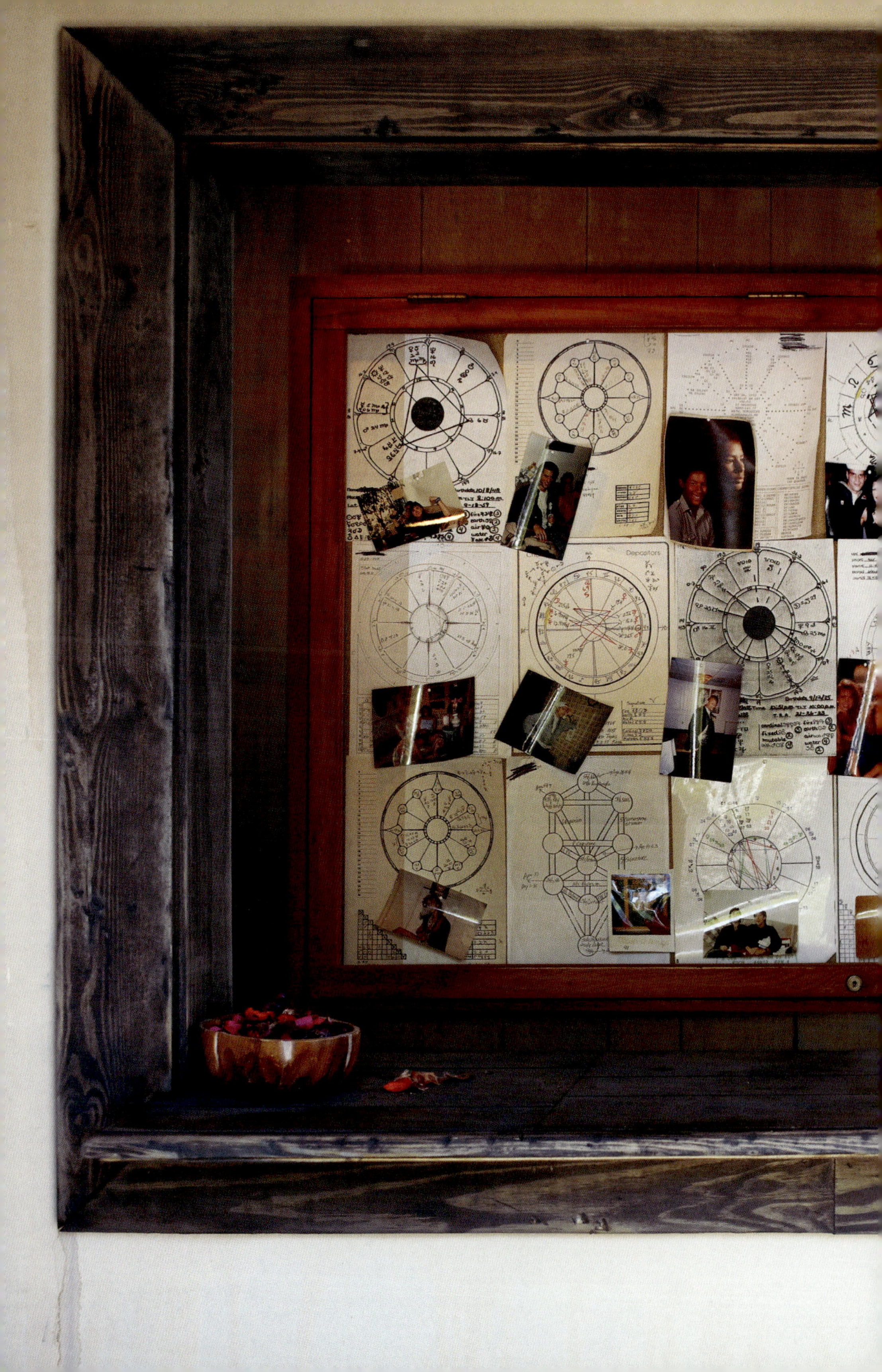
Depositors

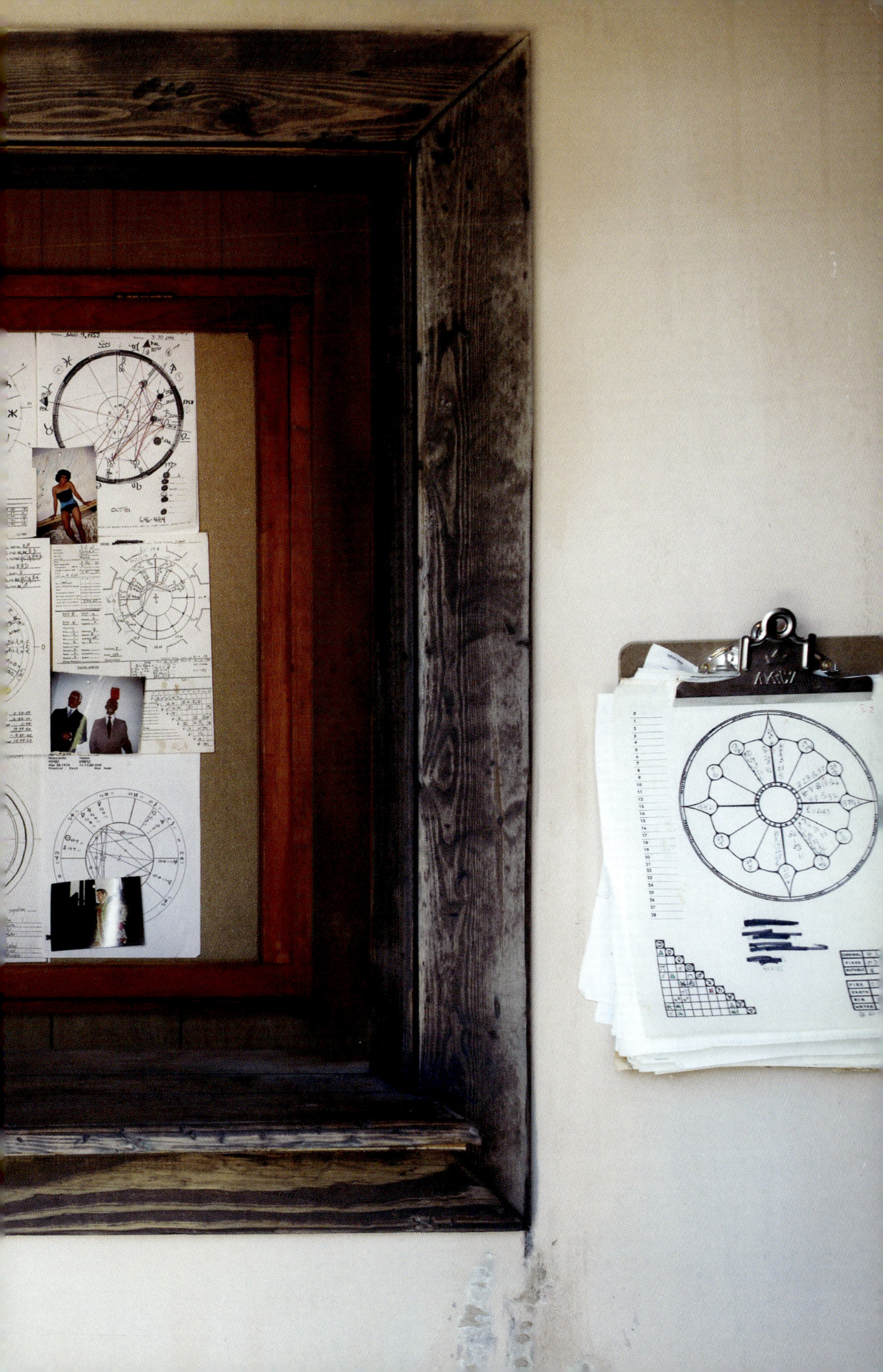

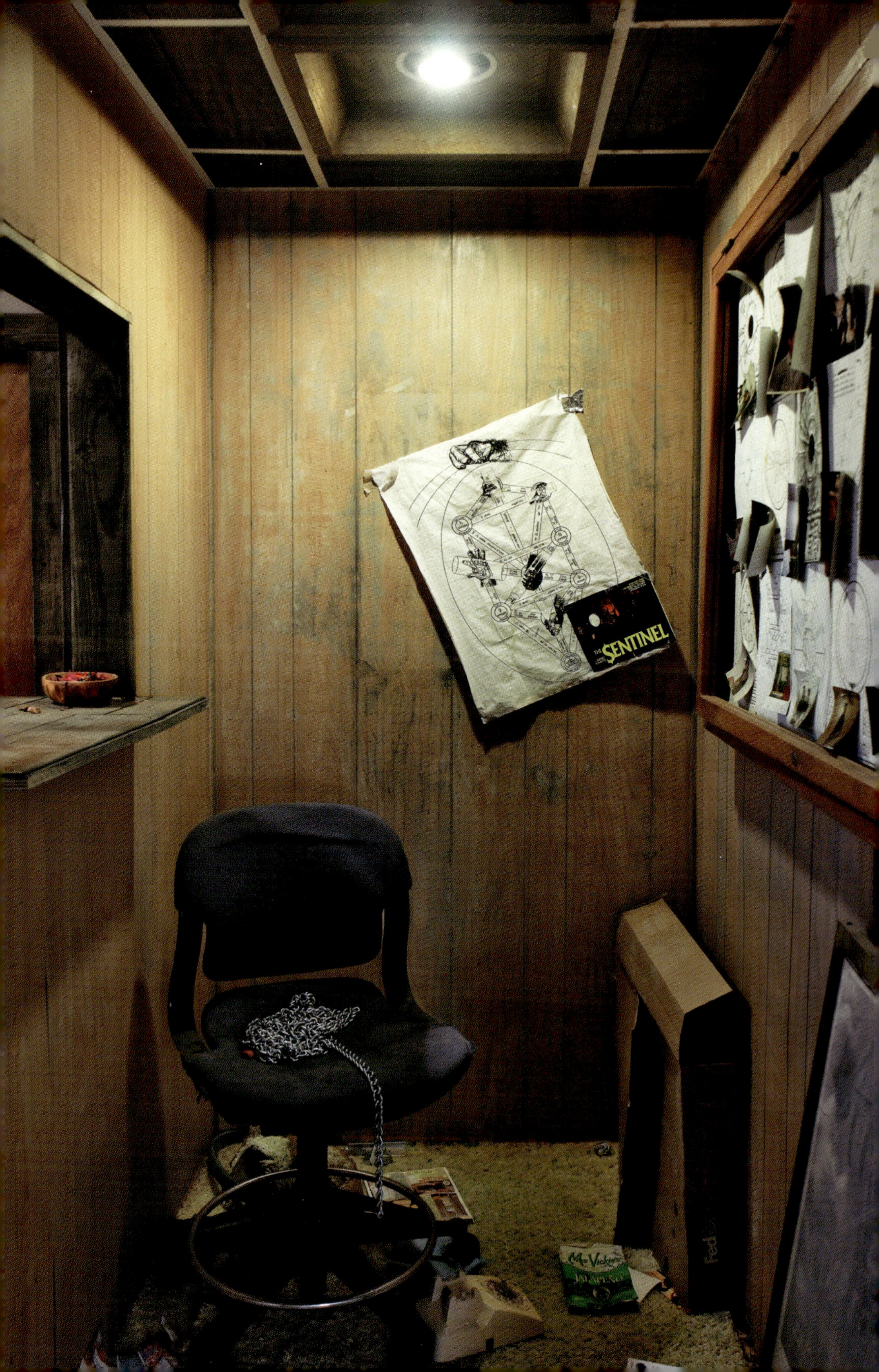
THE SENTINEL
Mr Vickie's

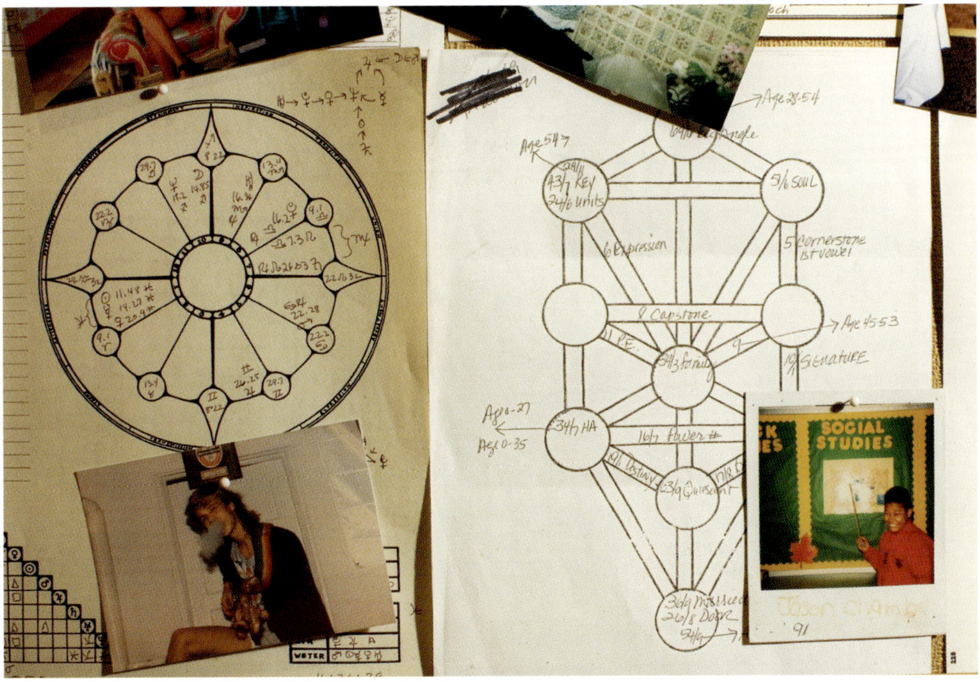
Capstone
SIGNATURE
SOCIAL STUDIES
91

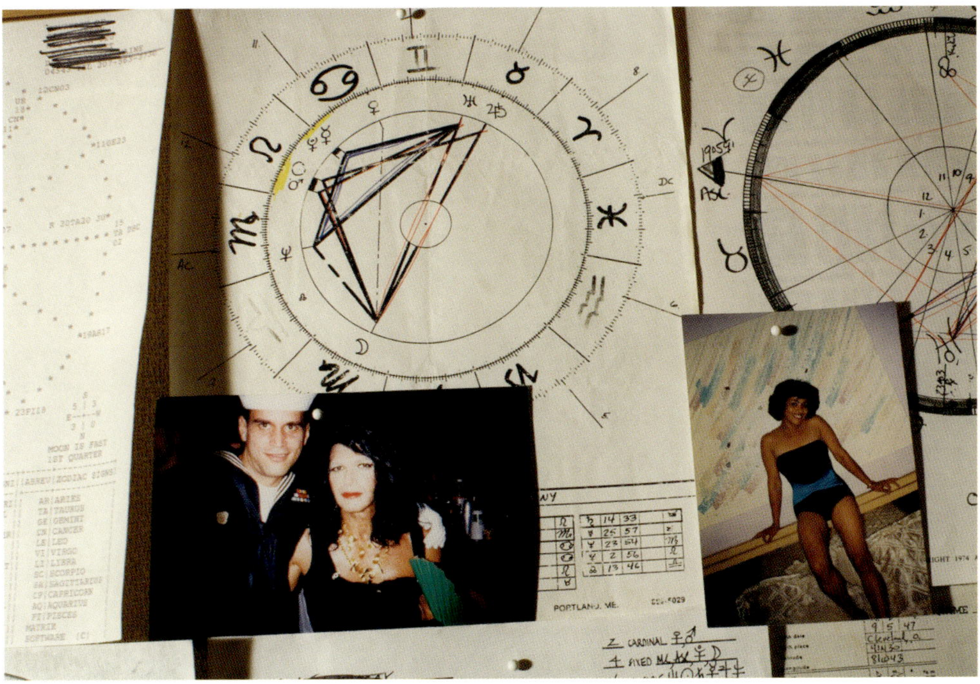
MOON IS PAST
1ST QUARTER
AR ARIES
TA TAURUS
GE GEMINI
CN CANCER
LE LEO
VI VIRGO
LI LIBRA
SC SCORPIO
SA SAGITTARIUS
CP CAPRICORN
AQ AQUARIUS
PI PISCES
MATRIX
SOFTWARE (C)
PORTLAND, ME.
CARDINAL
FIXED

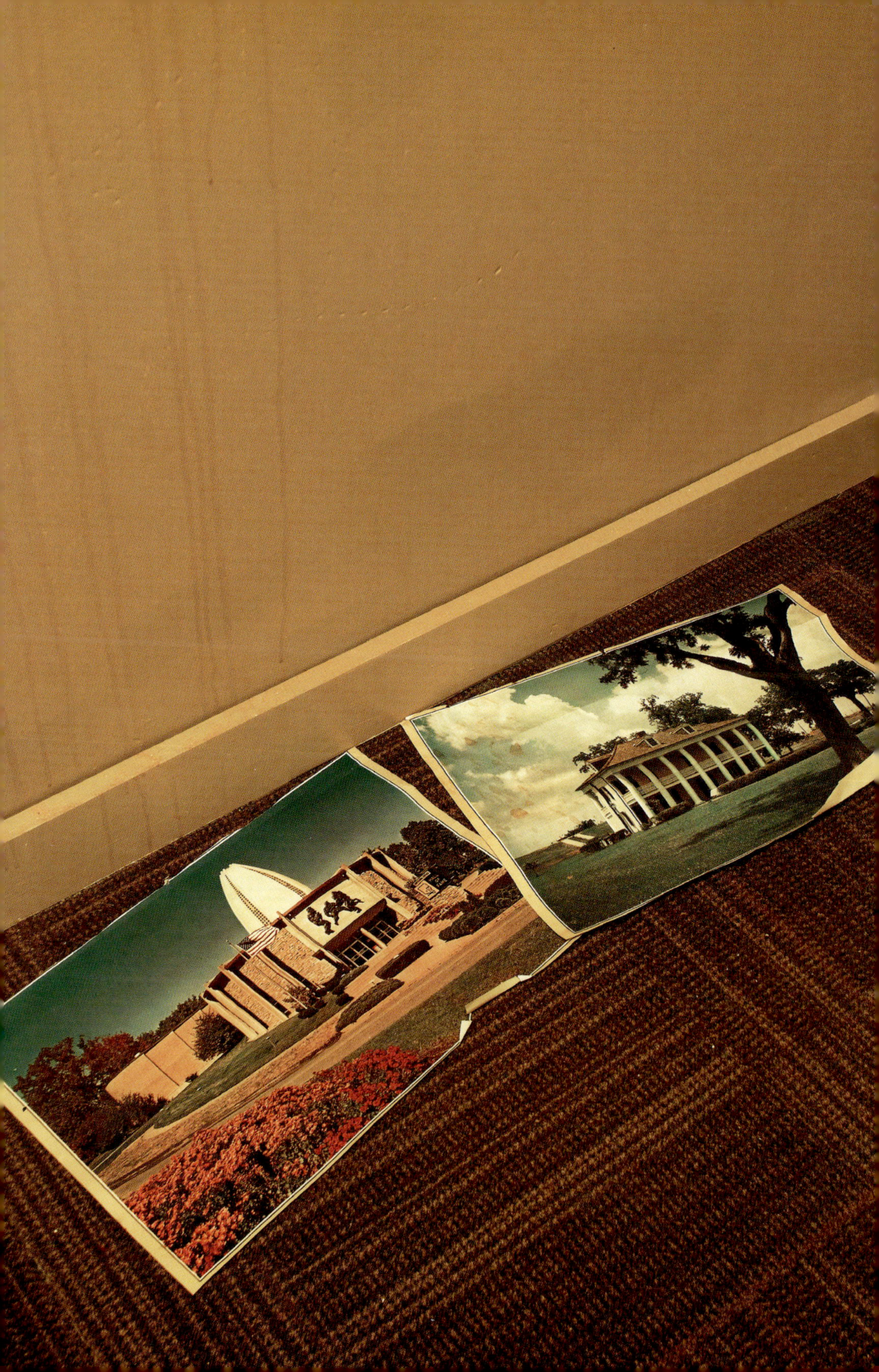

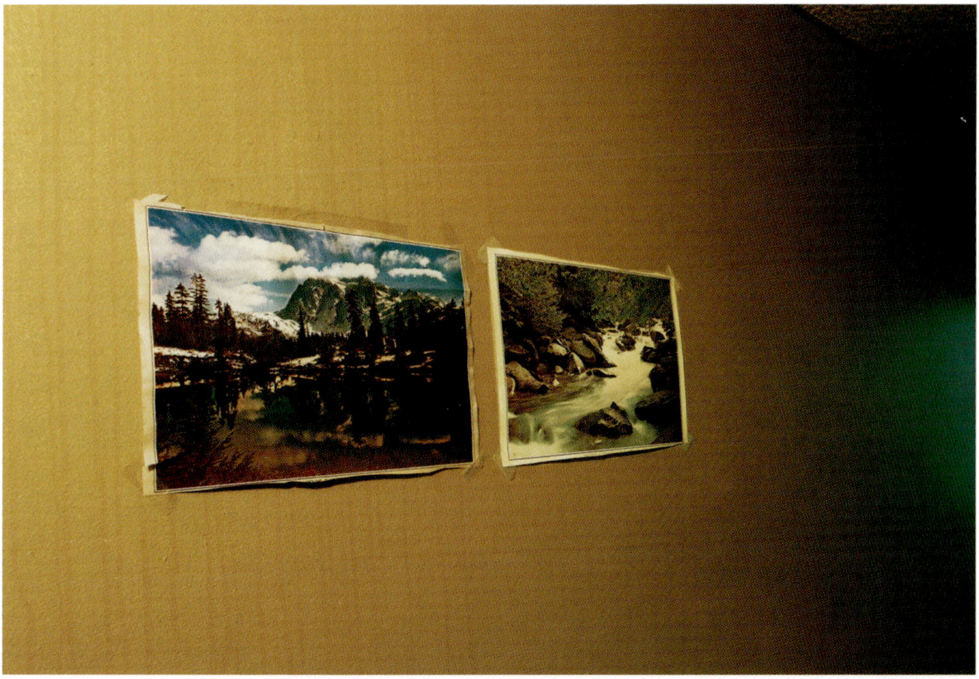

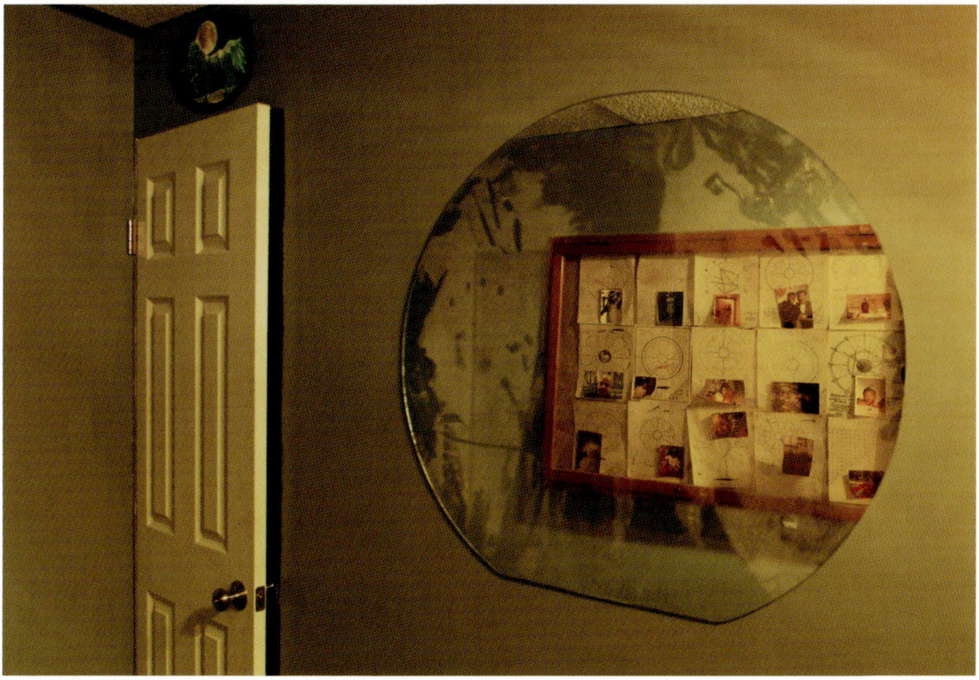

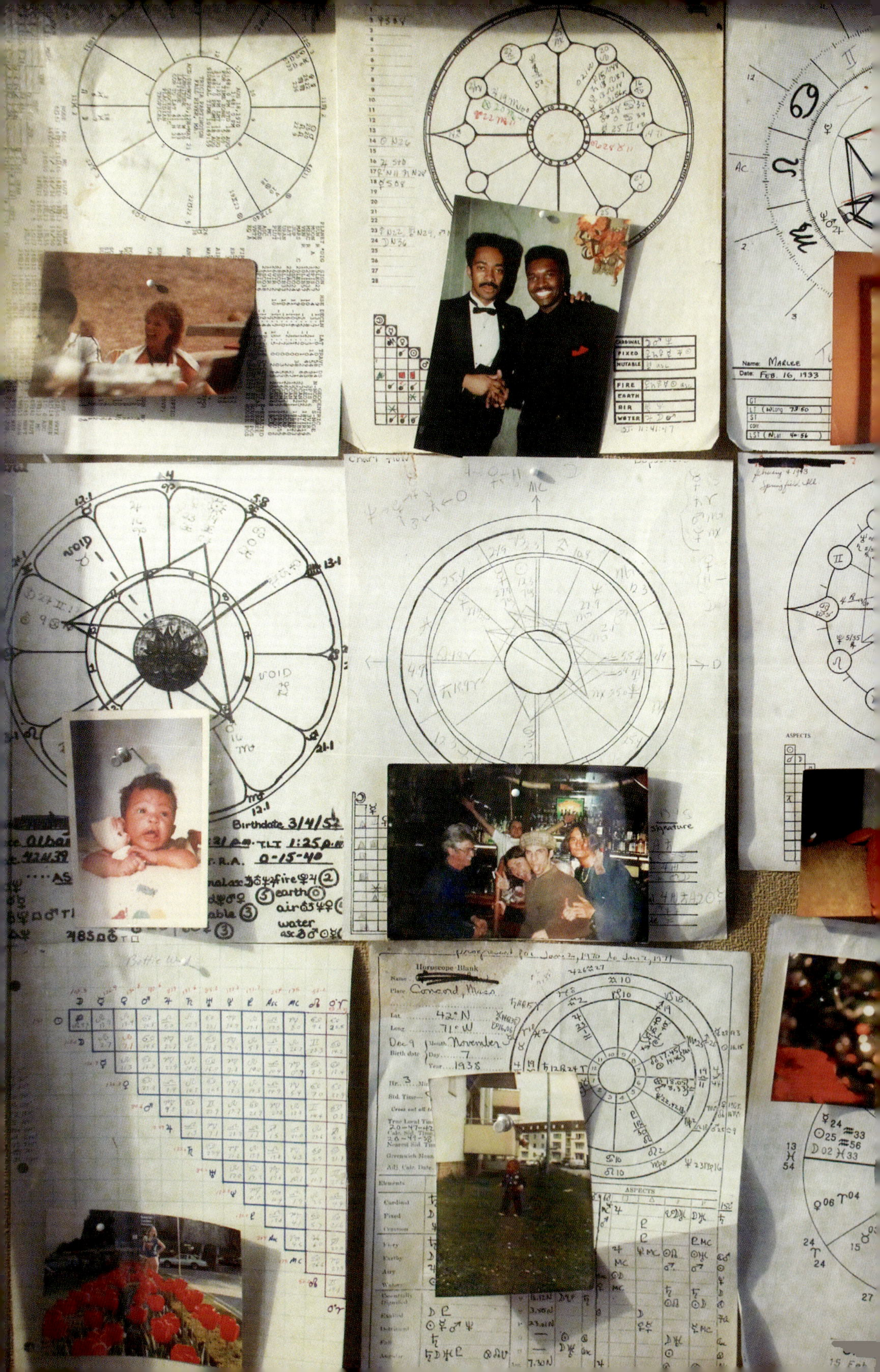

CARDINAL
FIXED
MUTABLE
FIRE
EARTH
AIR
WATER
Name: MARLEE
Date: FEB. 16, 1933
Birthdate 3/4/52
ASPECTS
Horoscope Blank
Concord, Mass.
November
1938

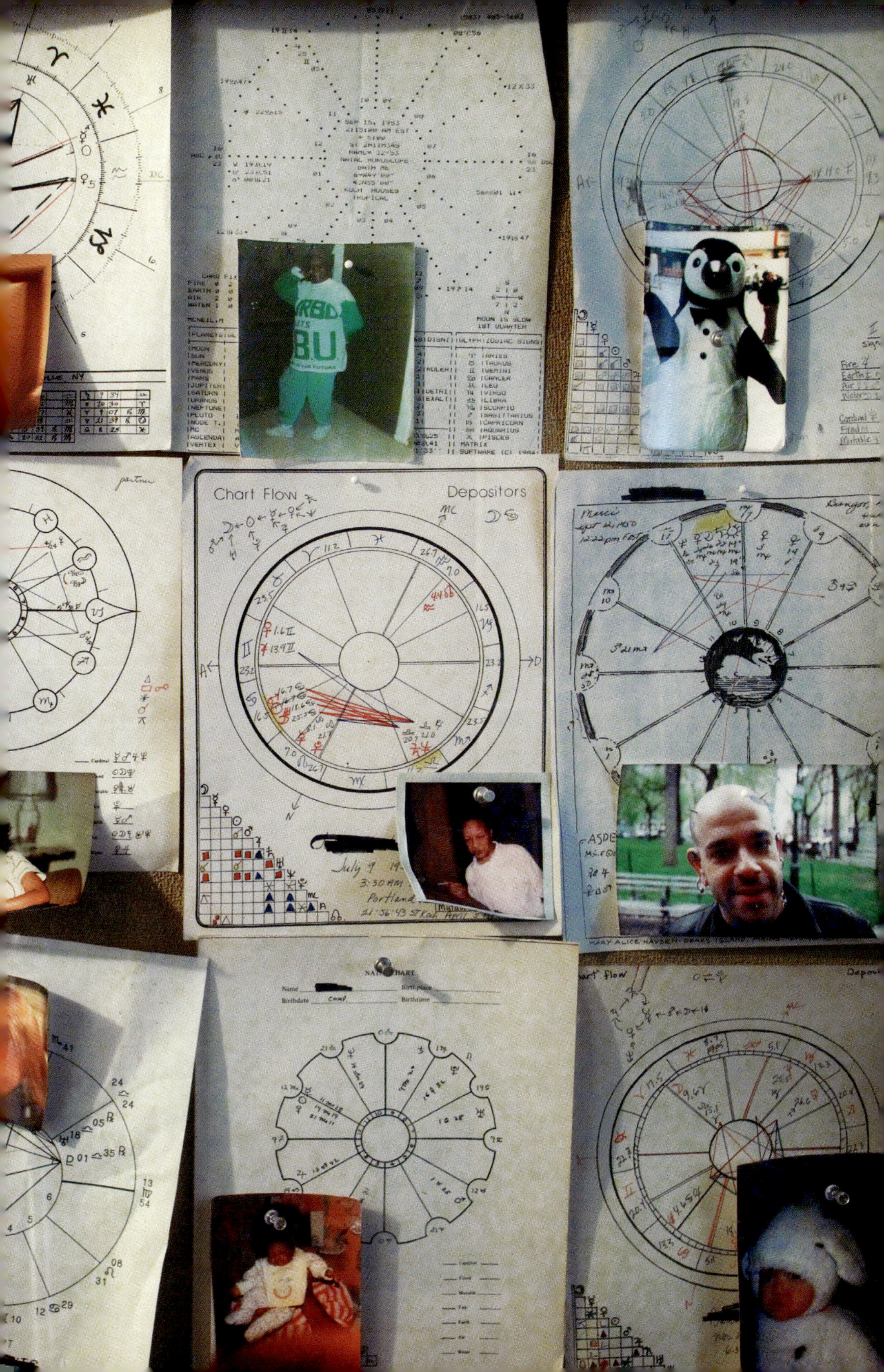
Chart Flow
Depositors

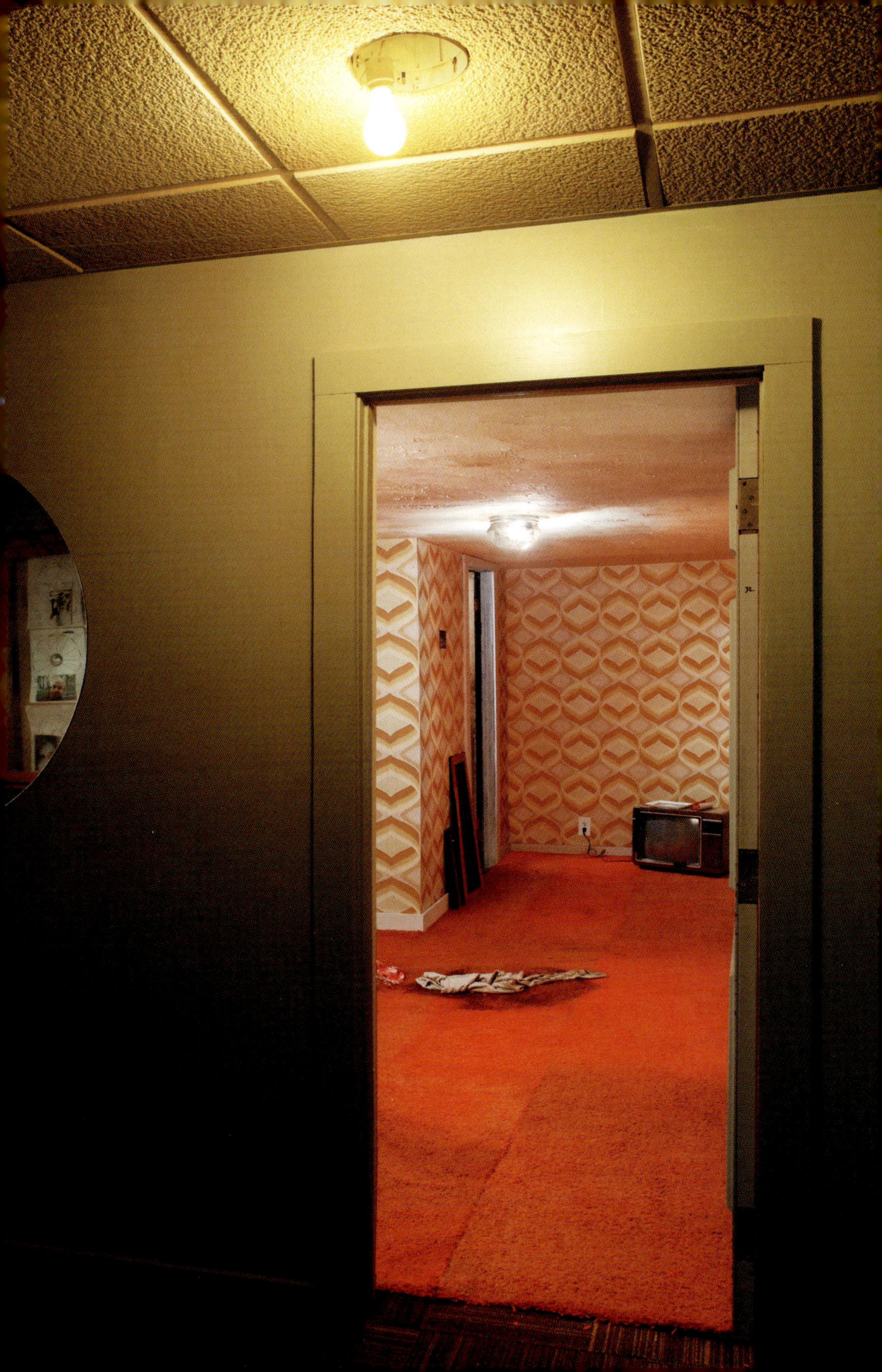

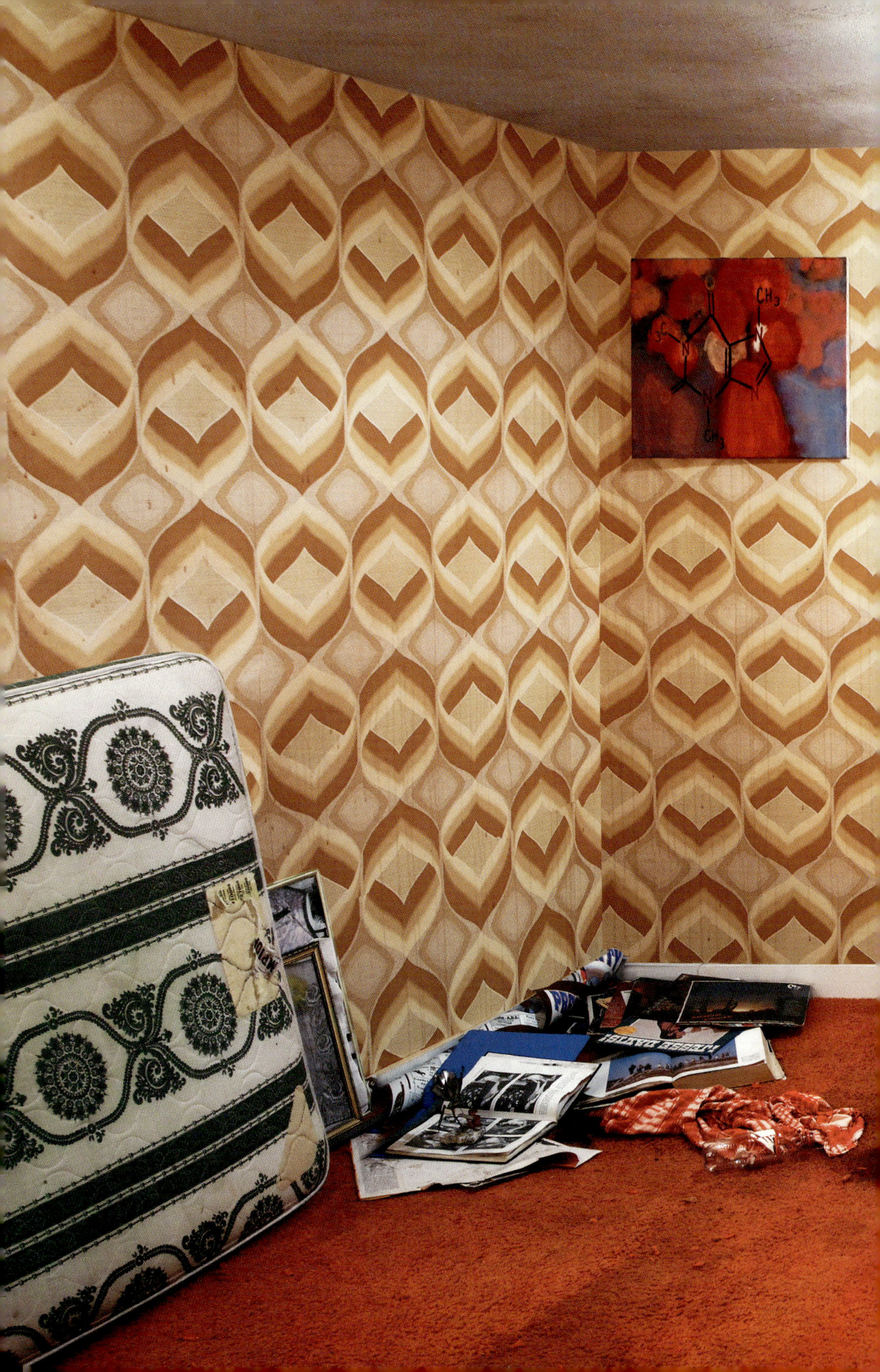

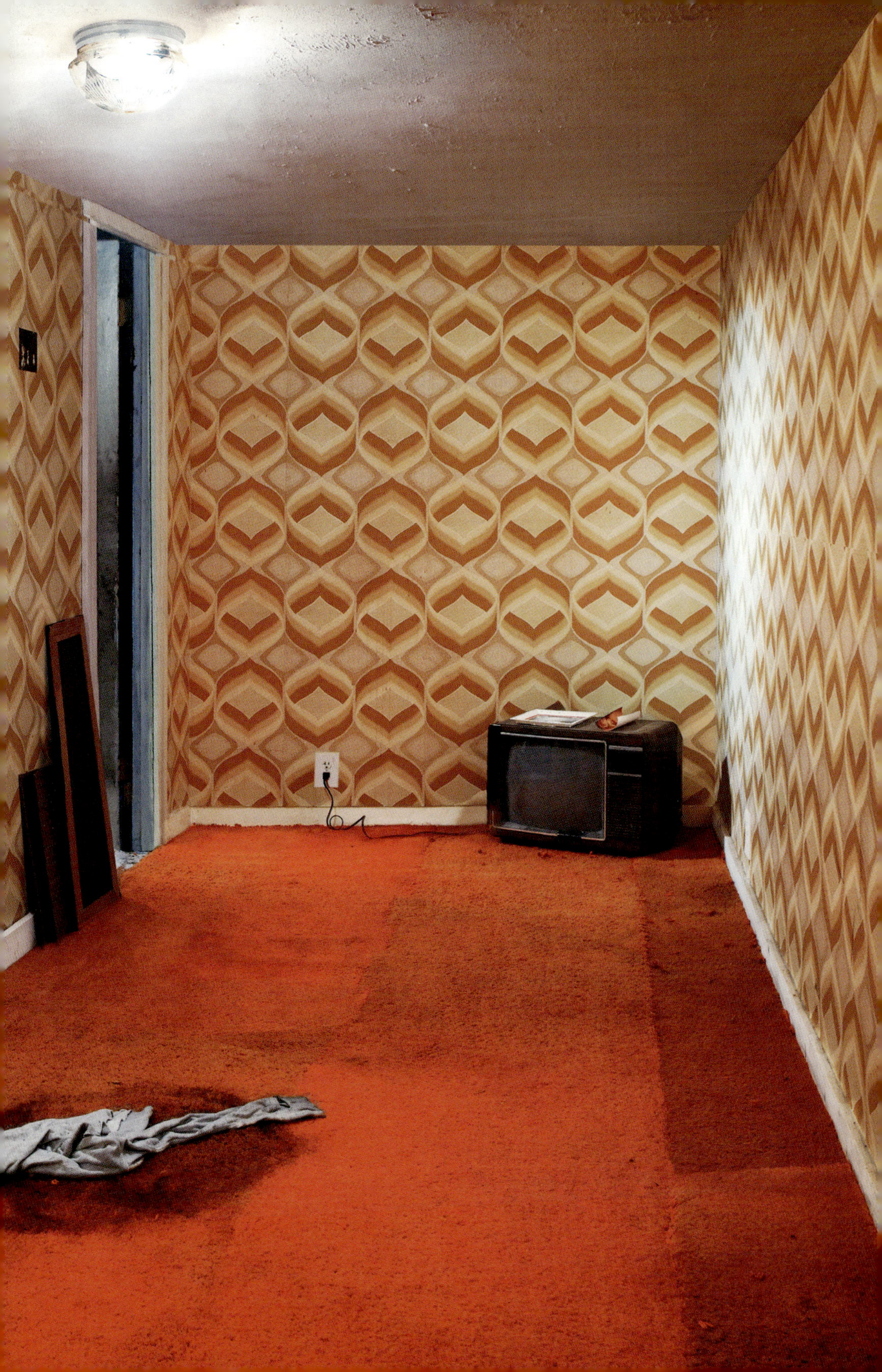

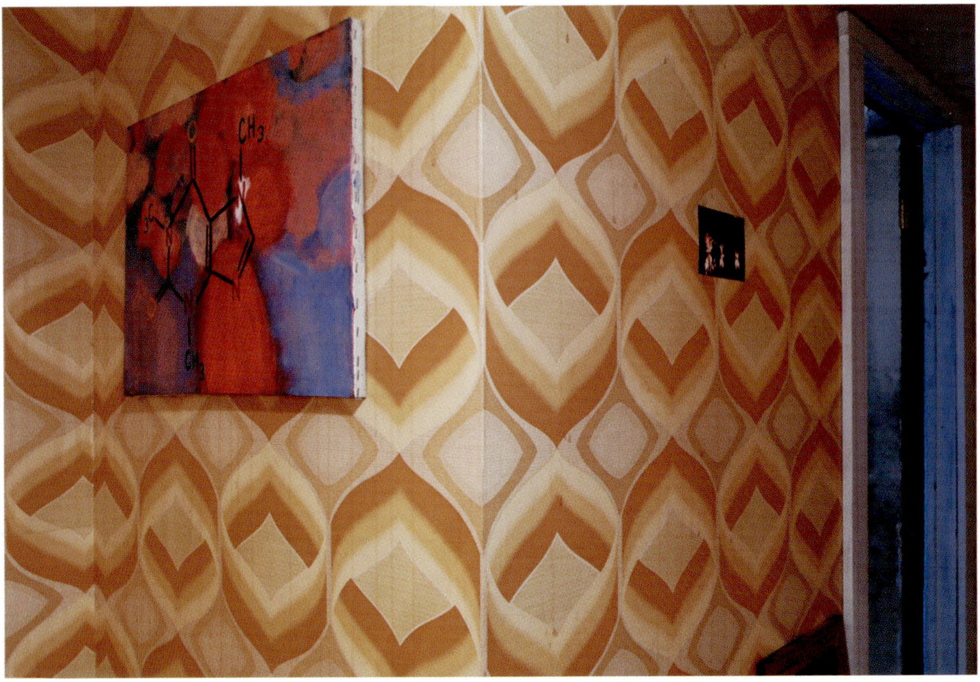
CH3

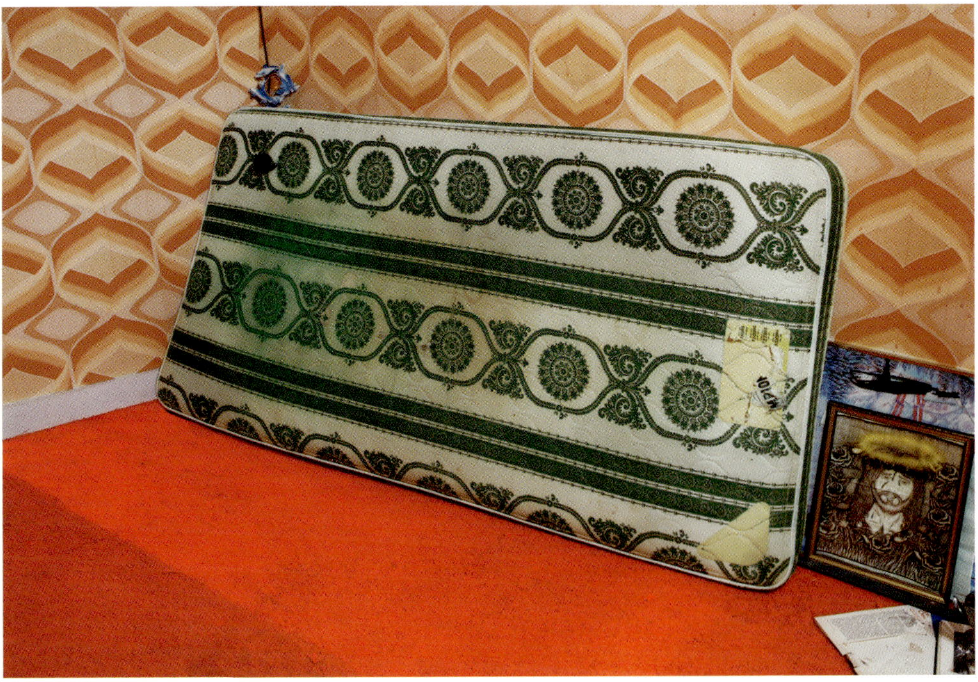

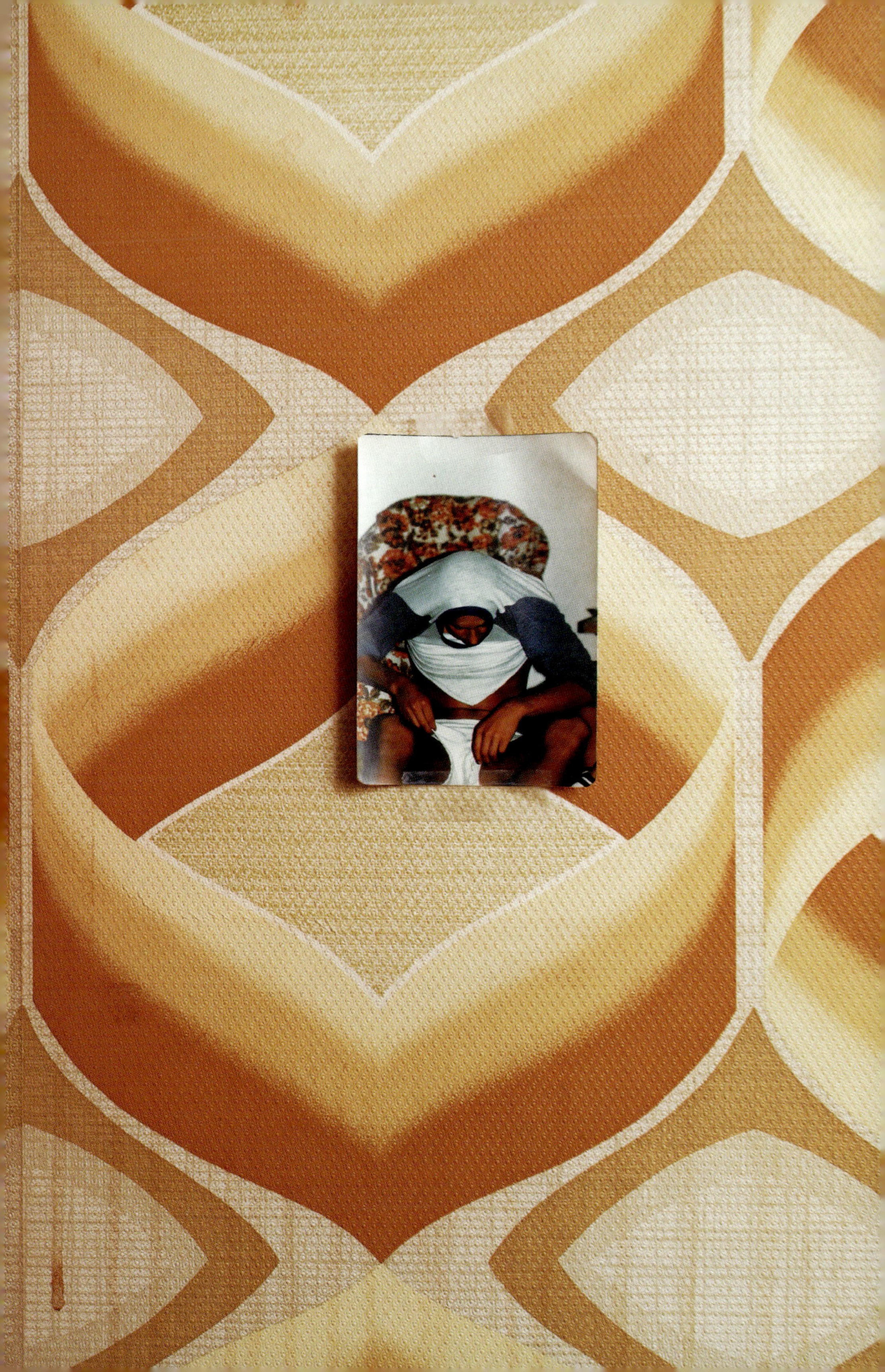

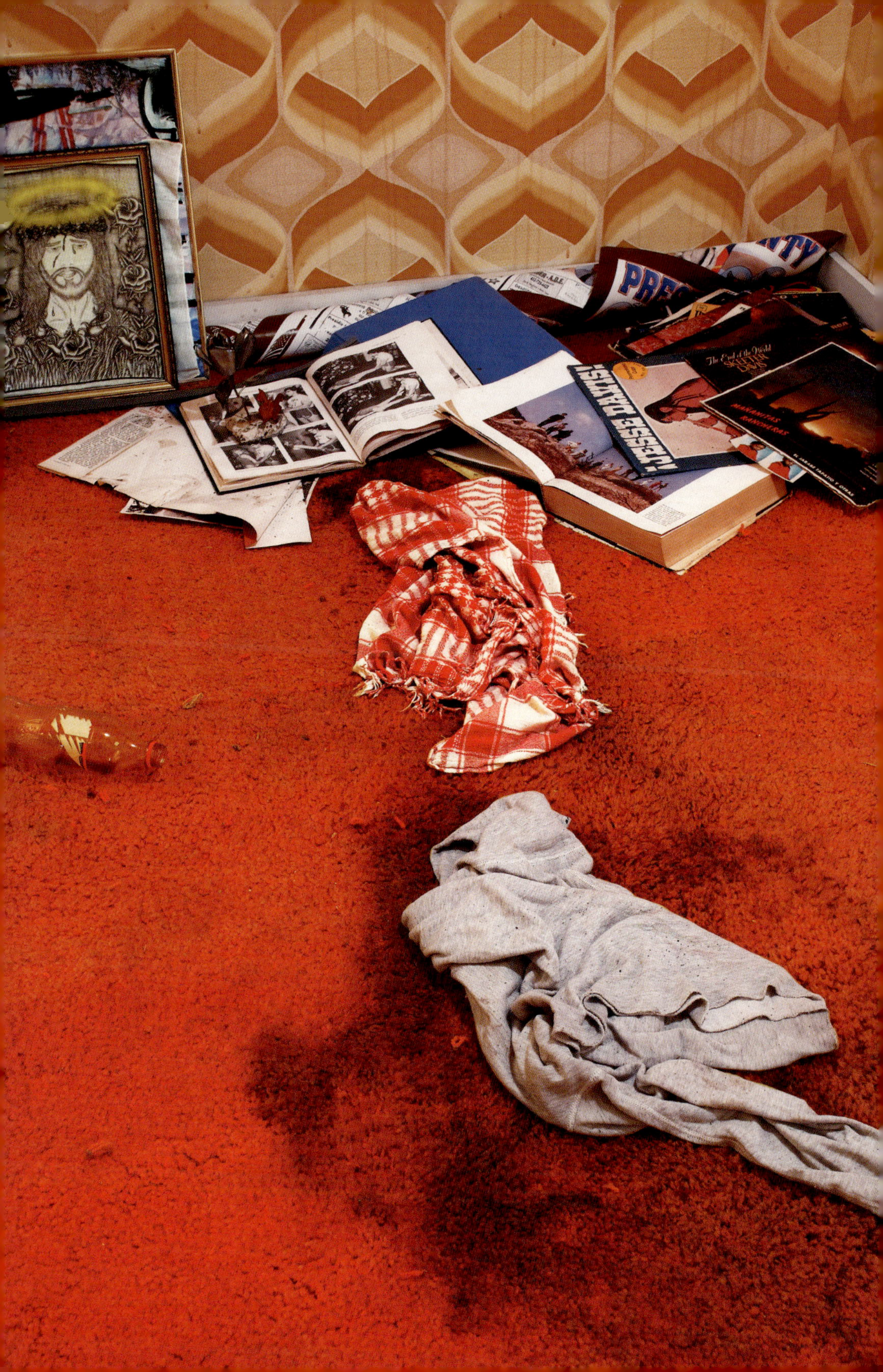

The End of the World
SKEETER DAVIS
MAÑANITAS RANCHERAS
EL JARABE TAPATIO Y OTRAS

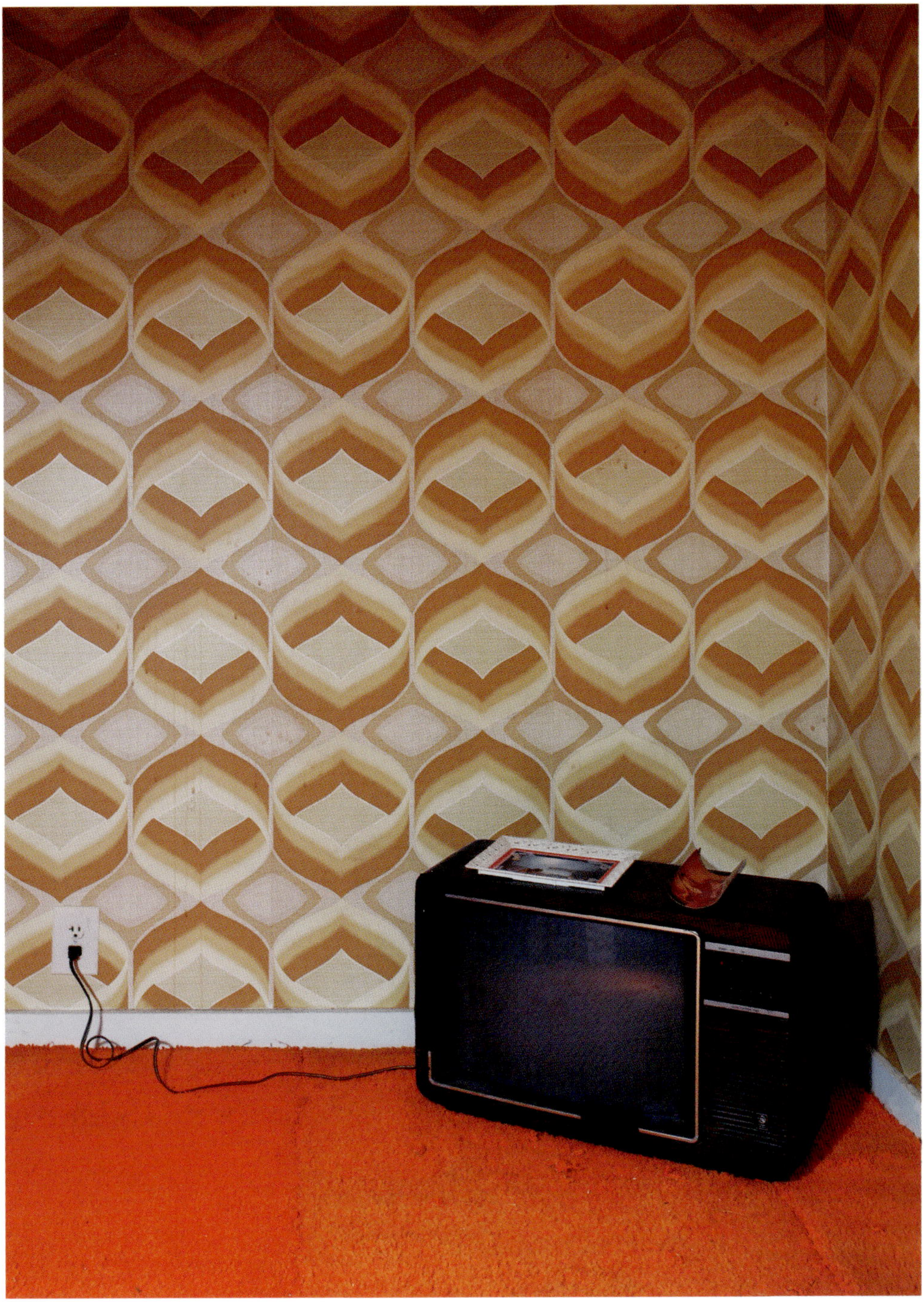

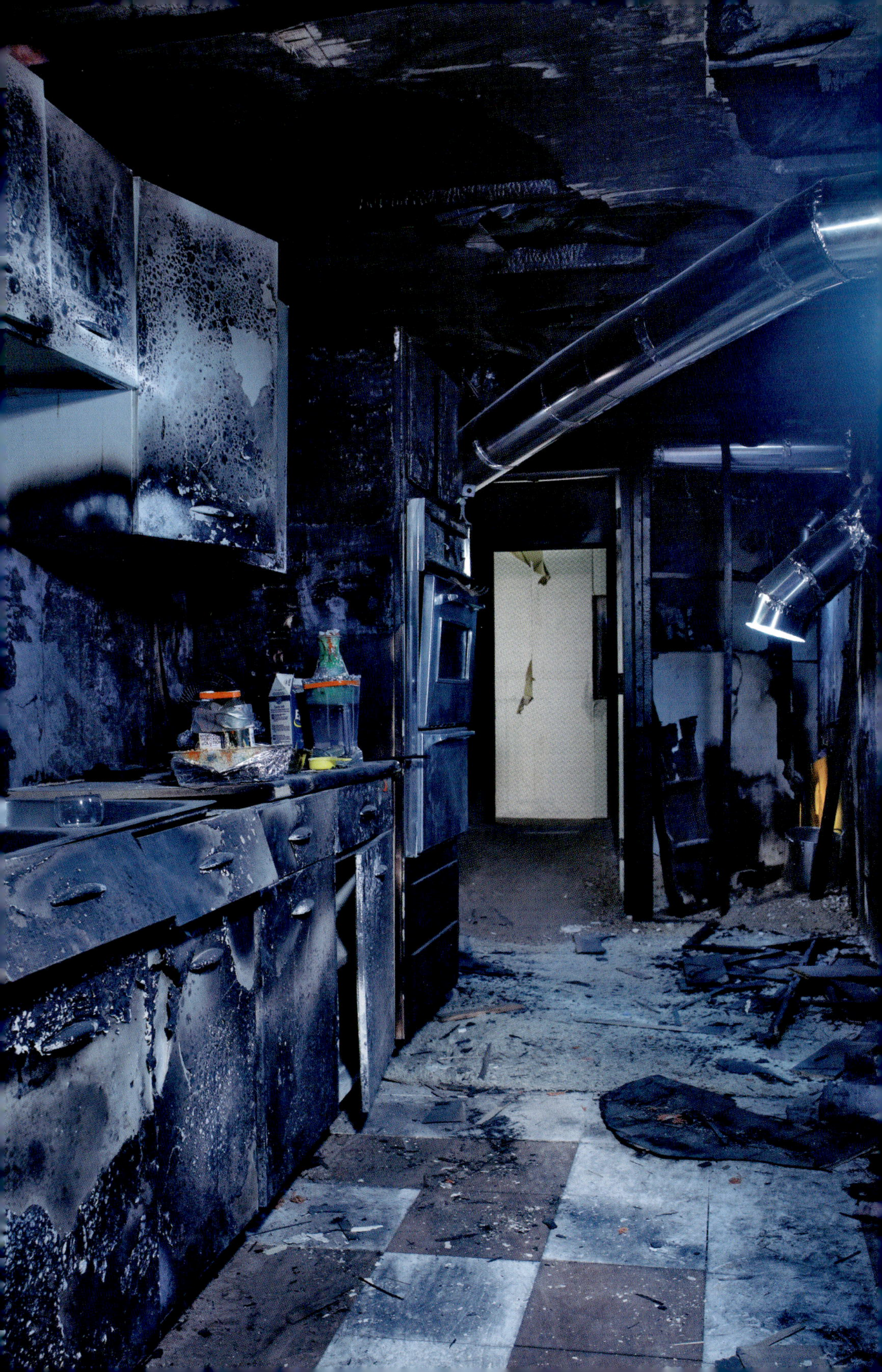

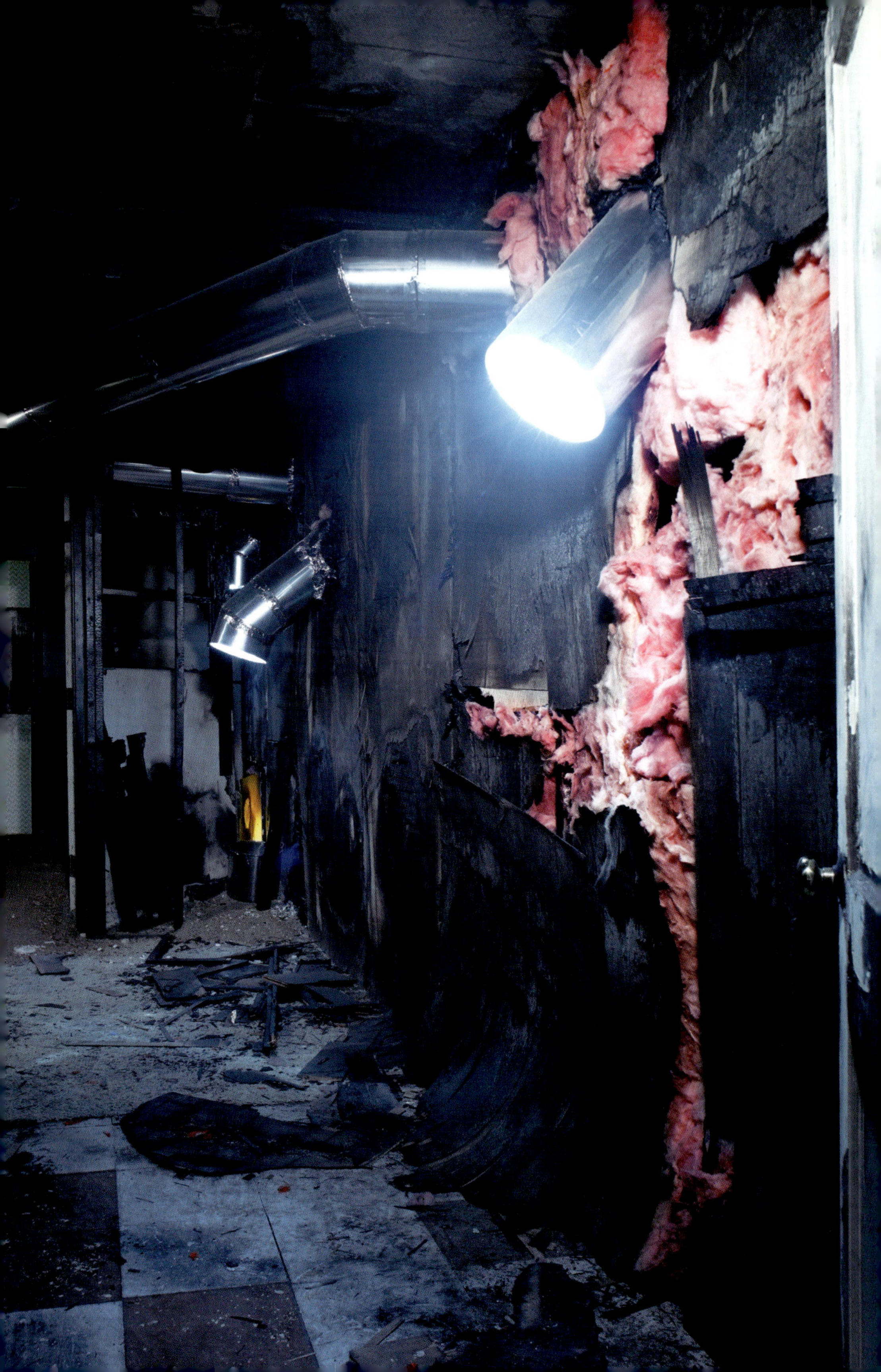

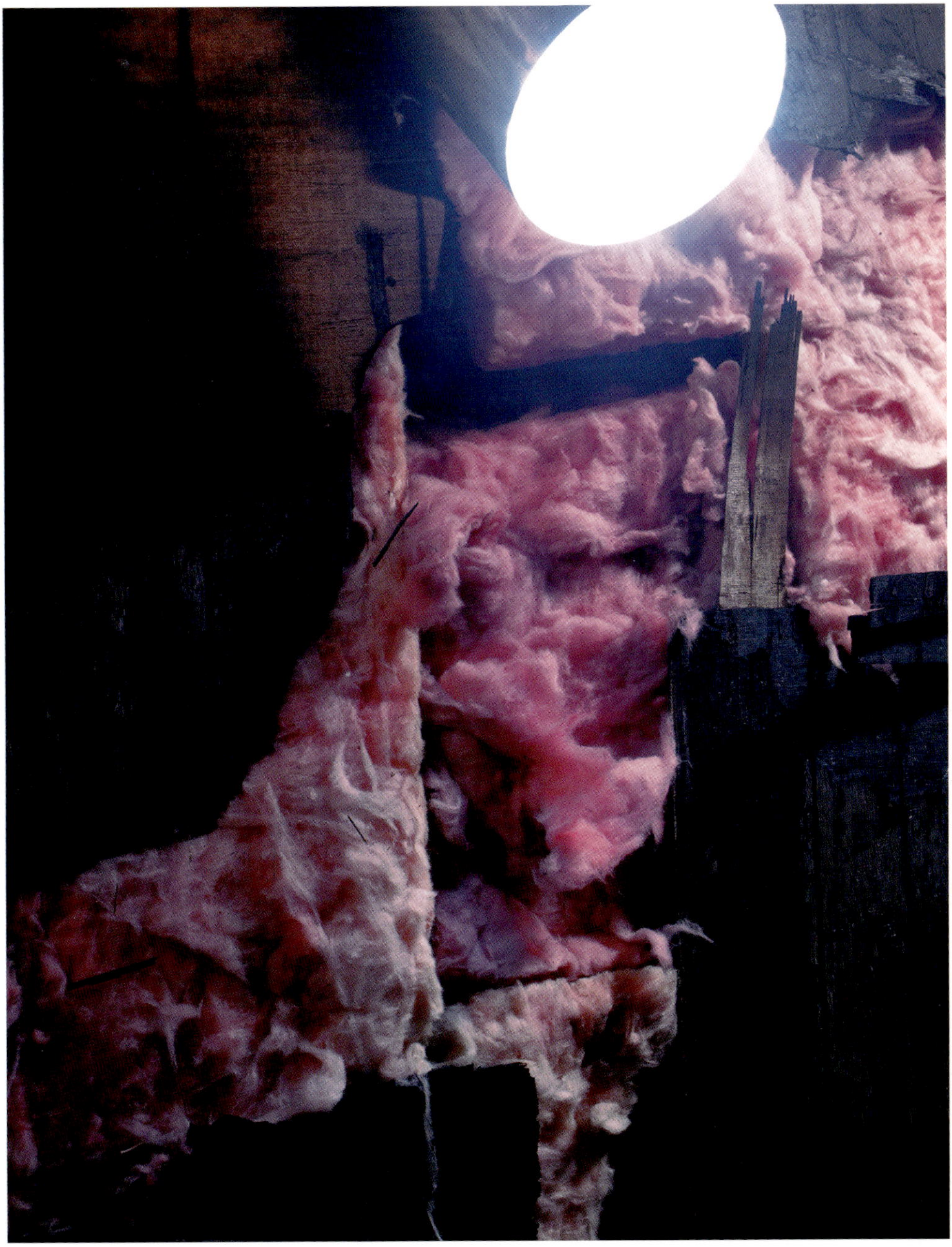

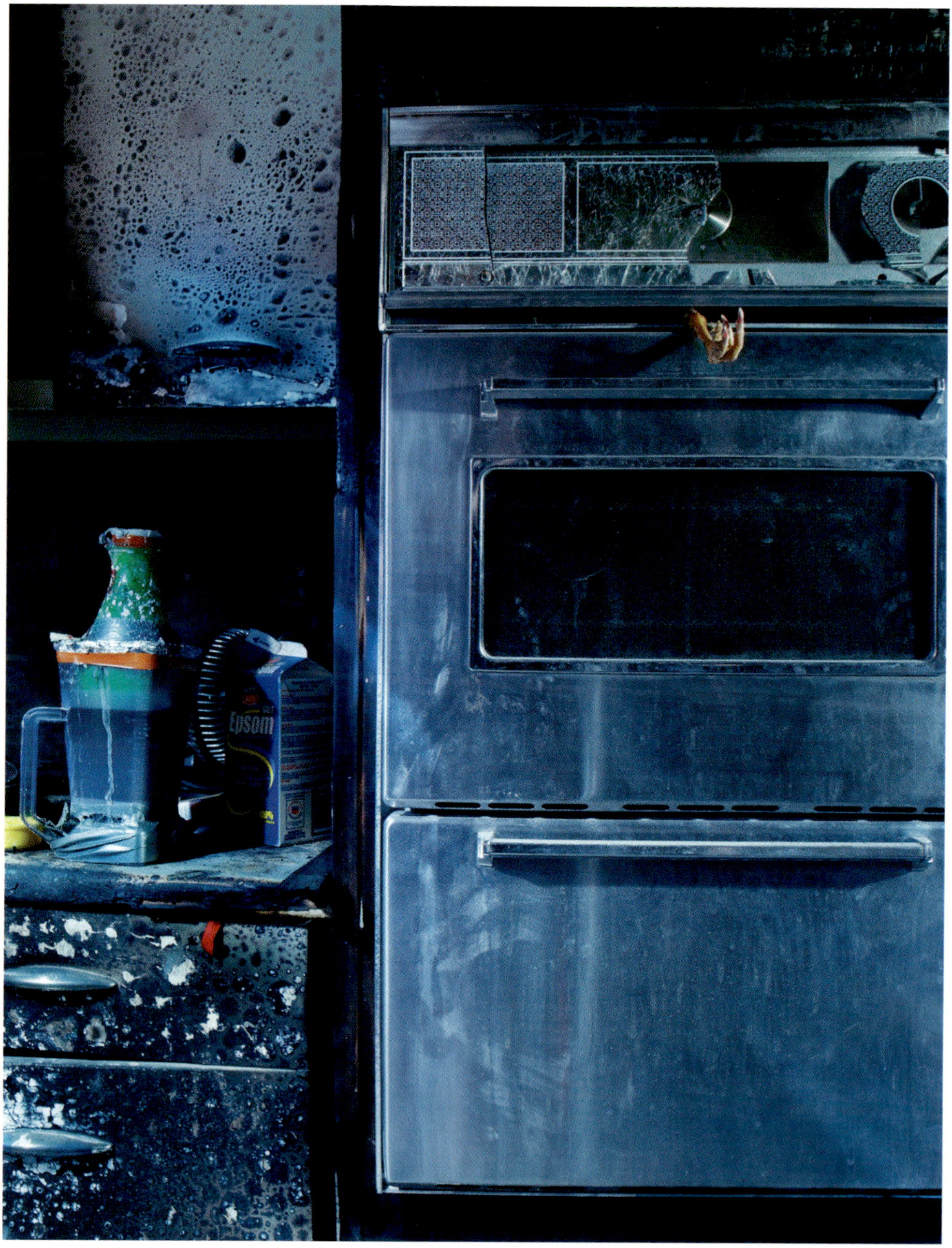
Epsom

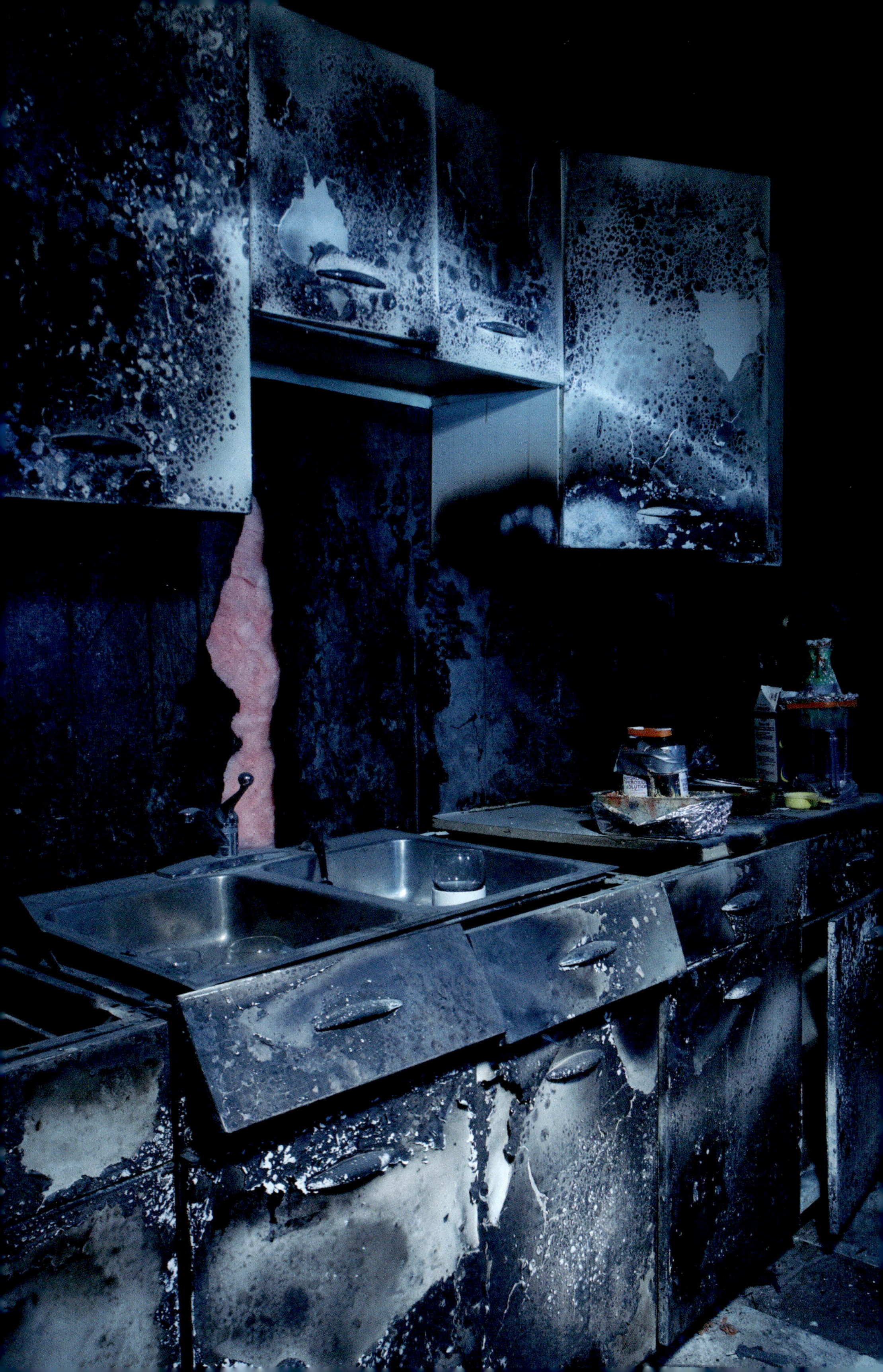

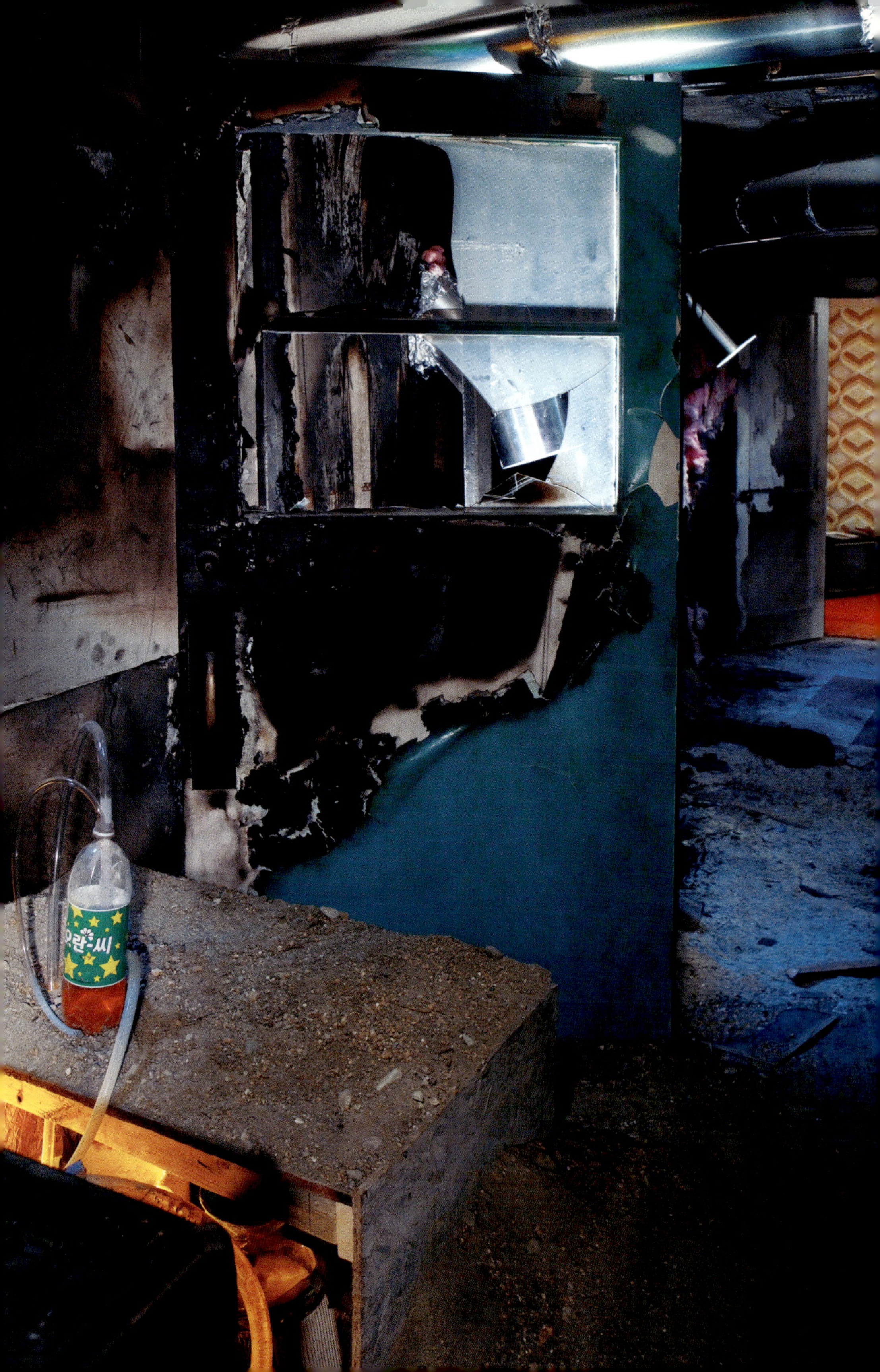
오란-씨

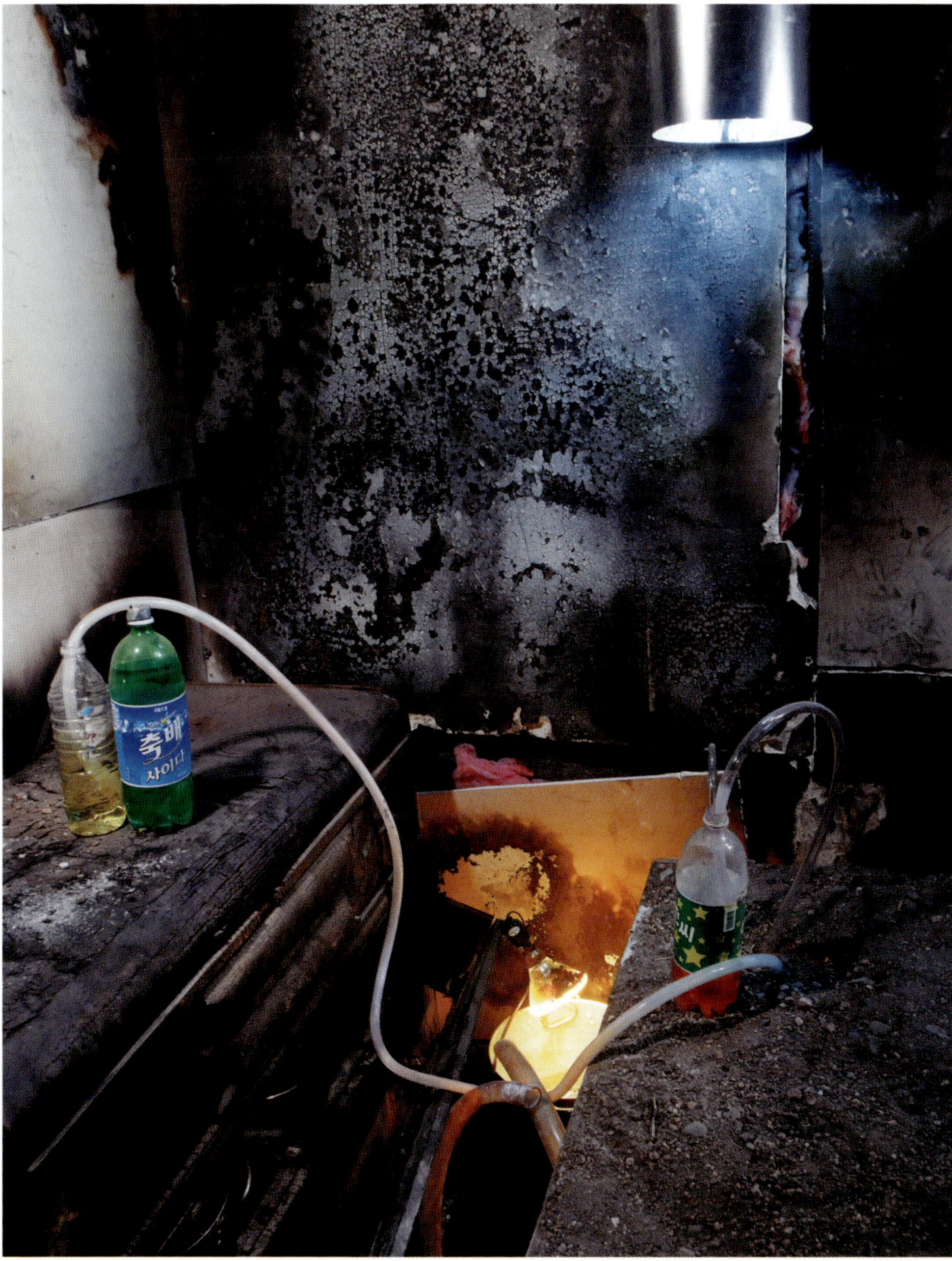
축배
사이다

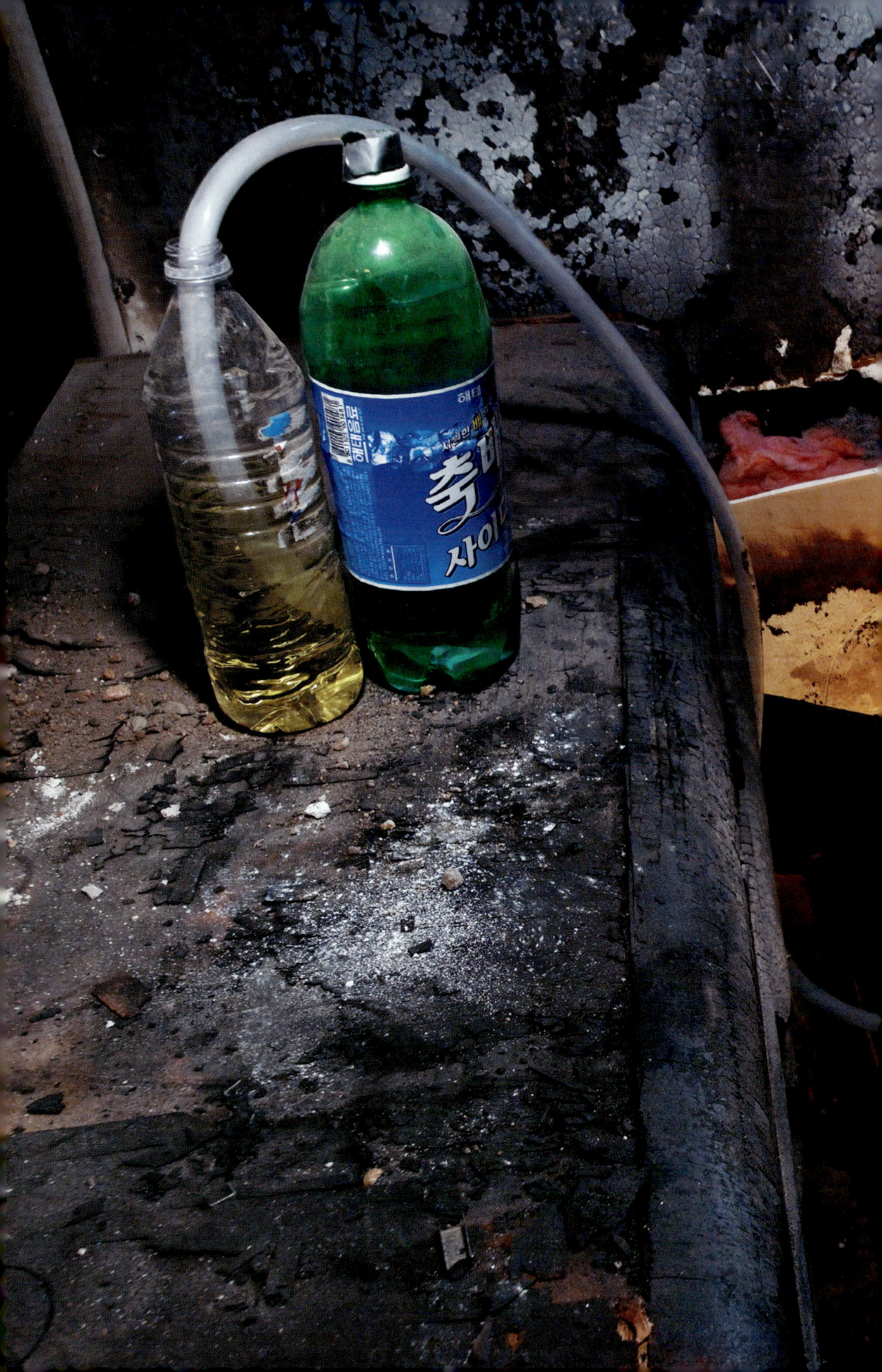
해태
축
사이

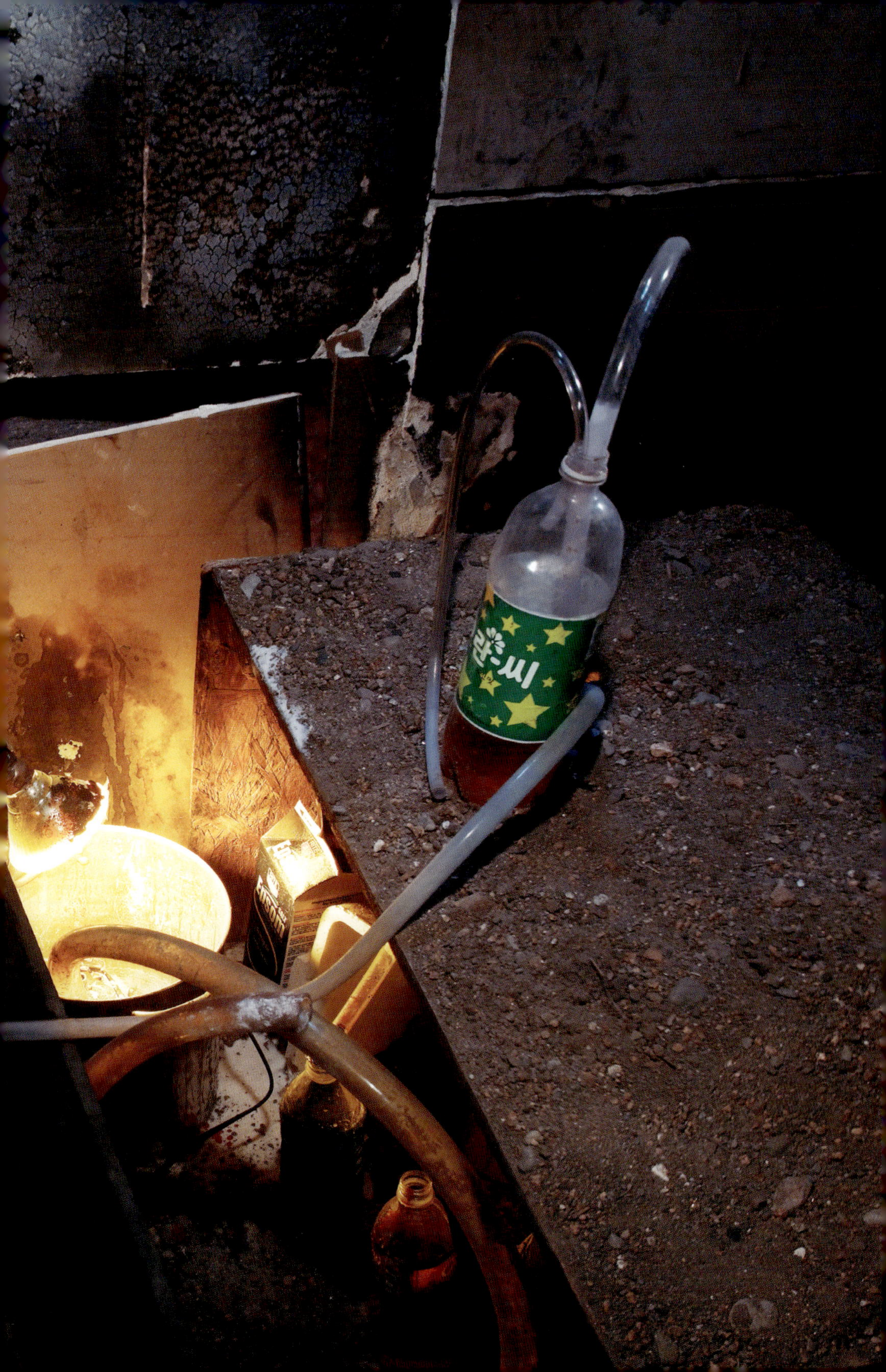

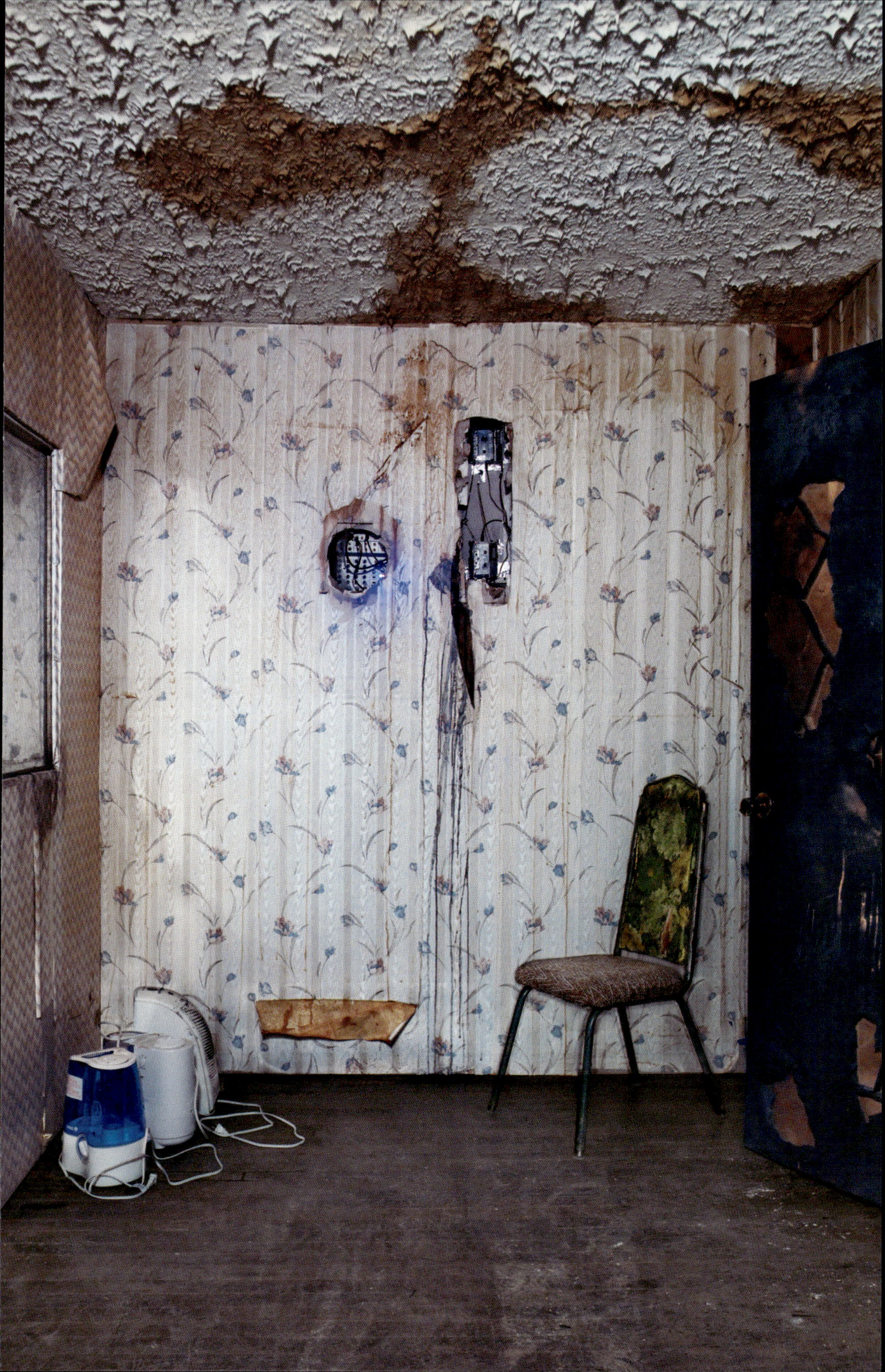

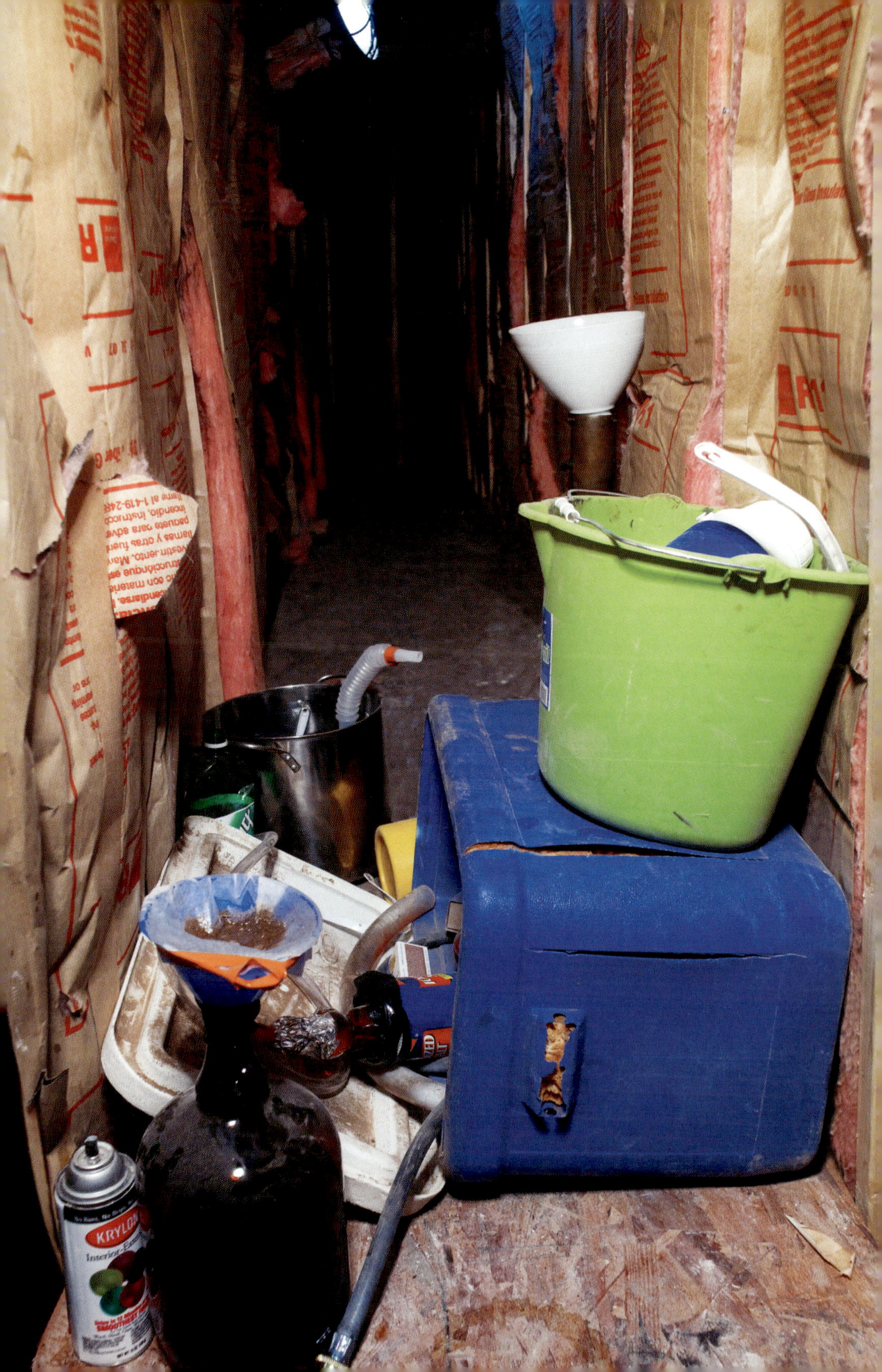
KRYLON

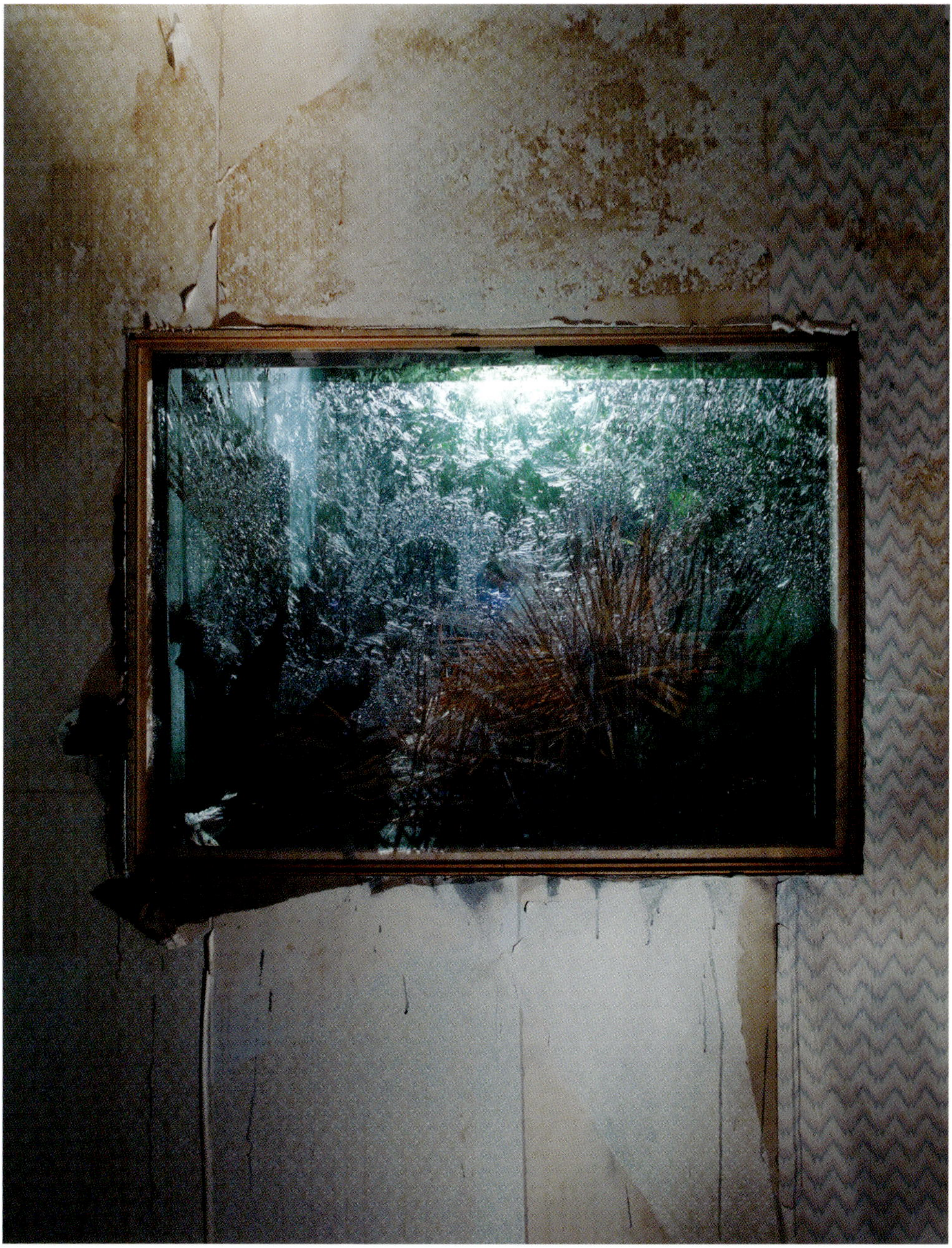

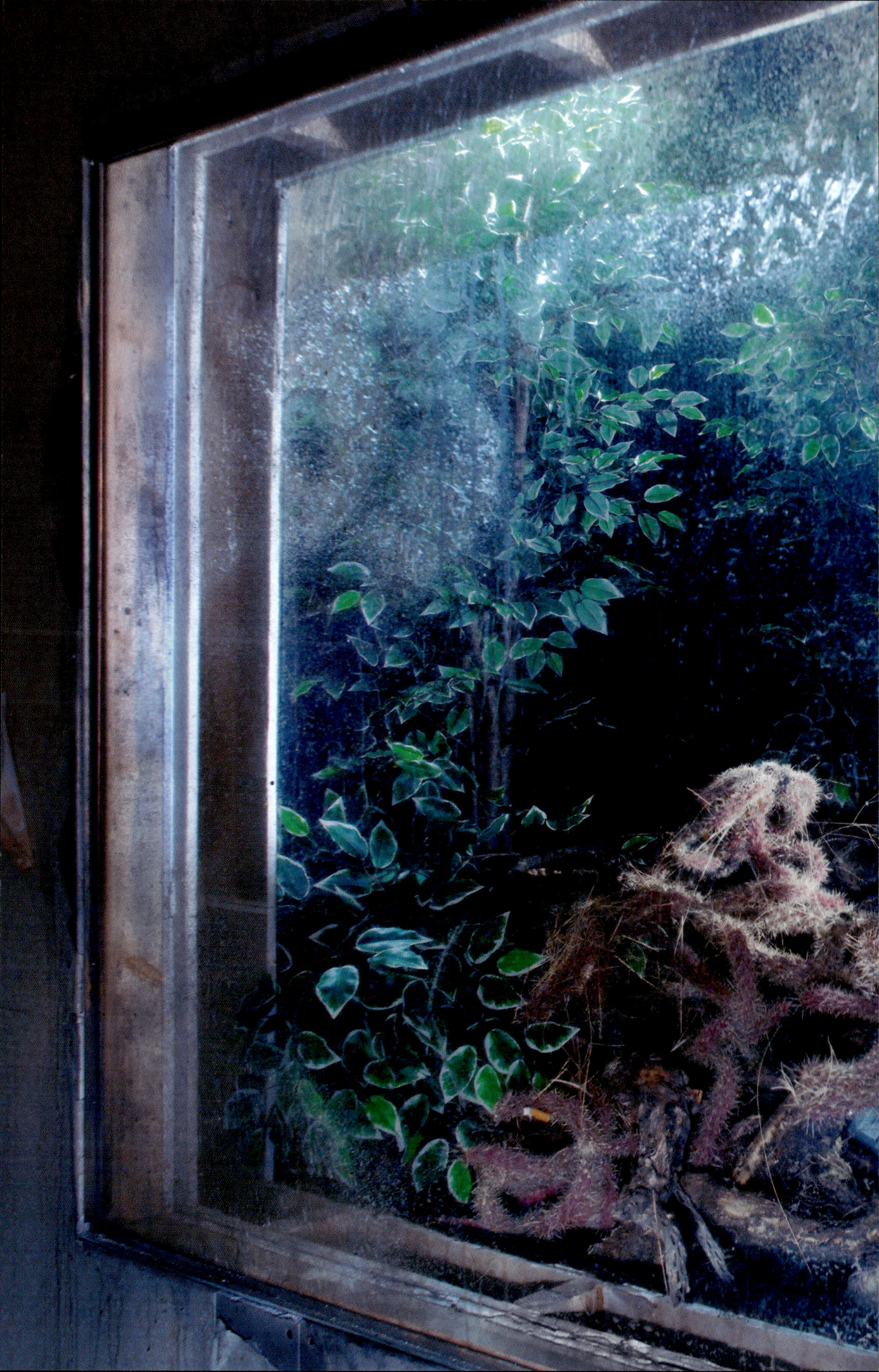

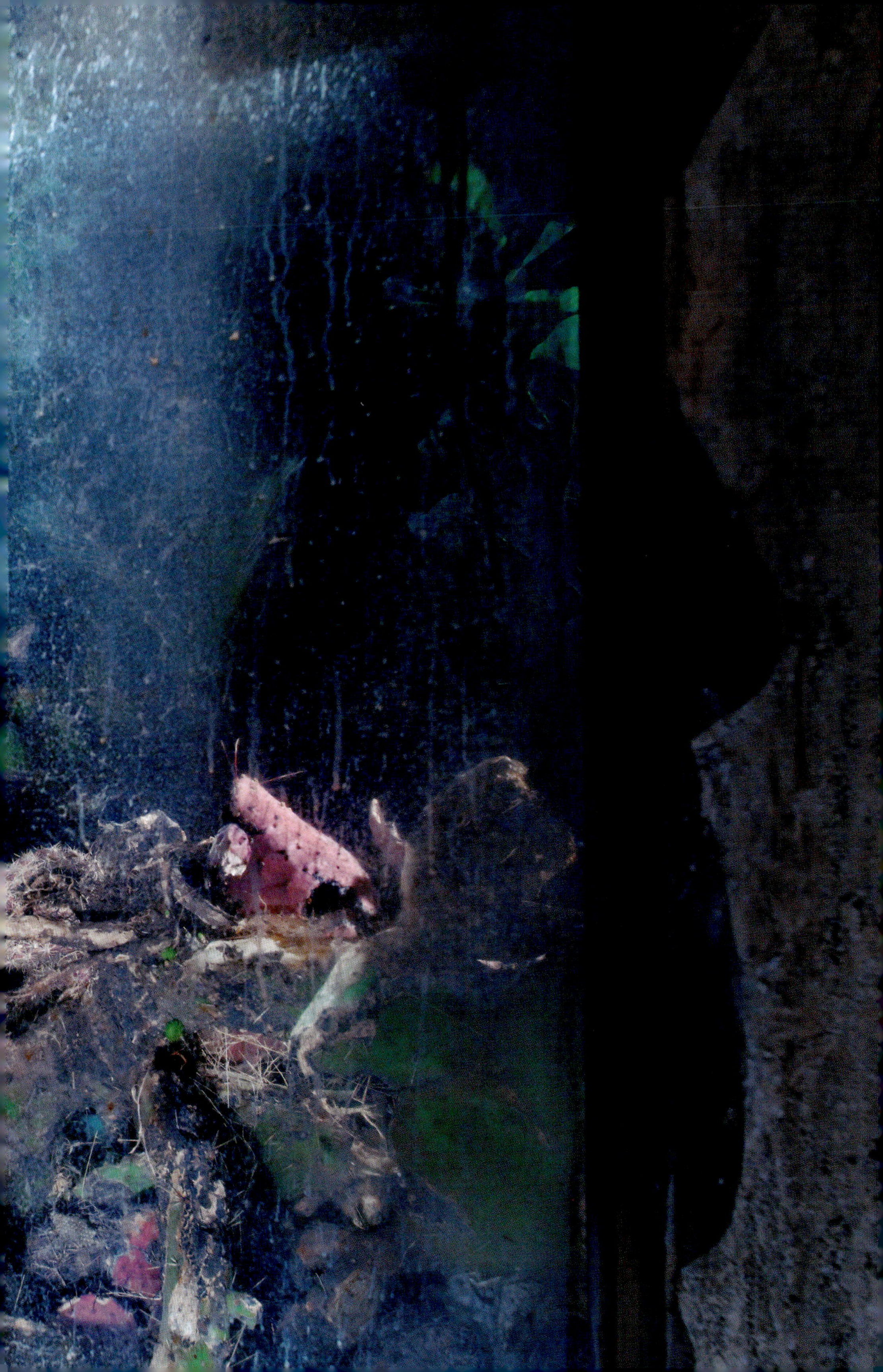

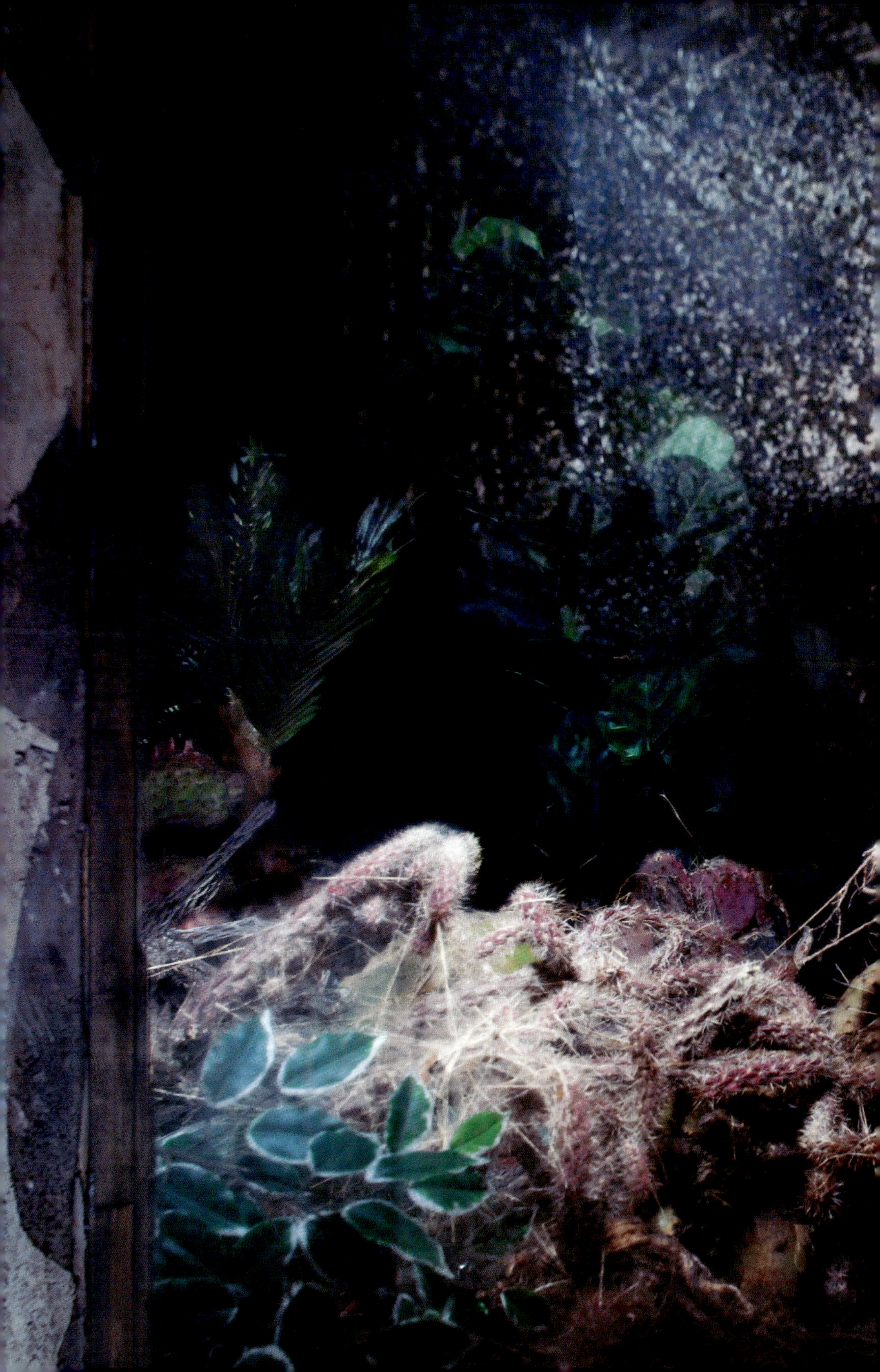

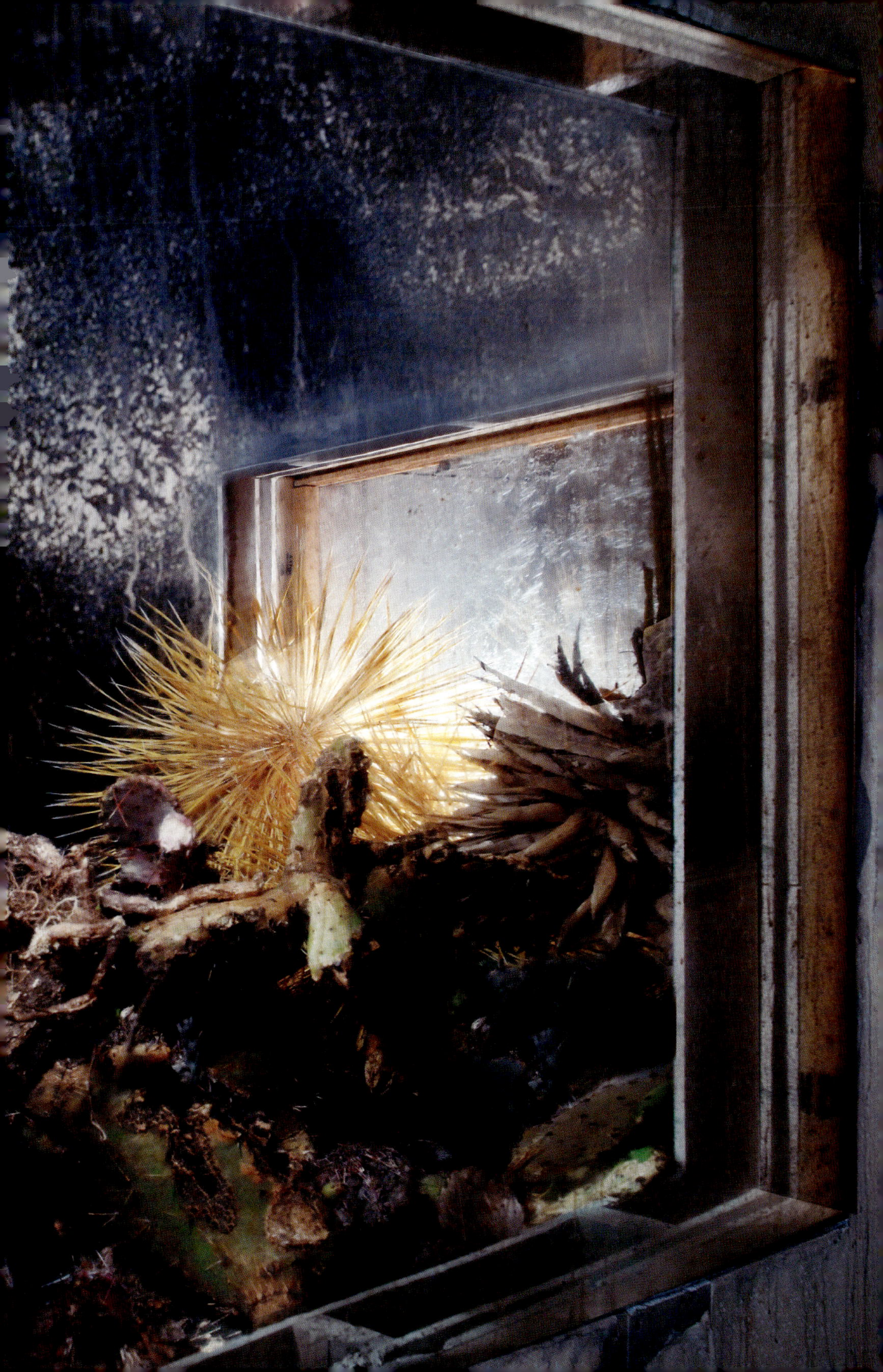

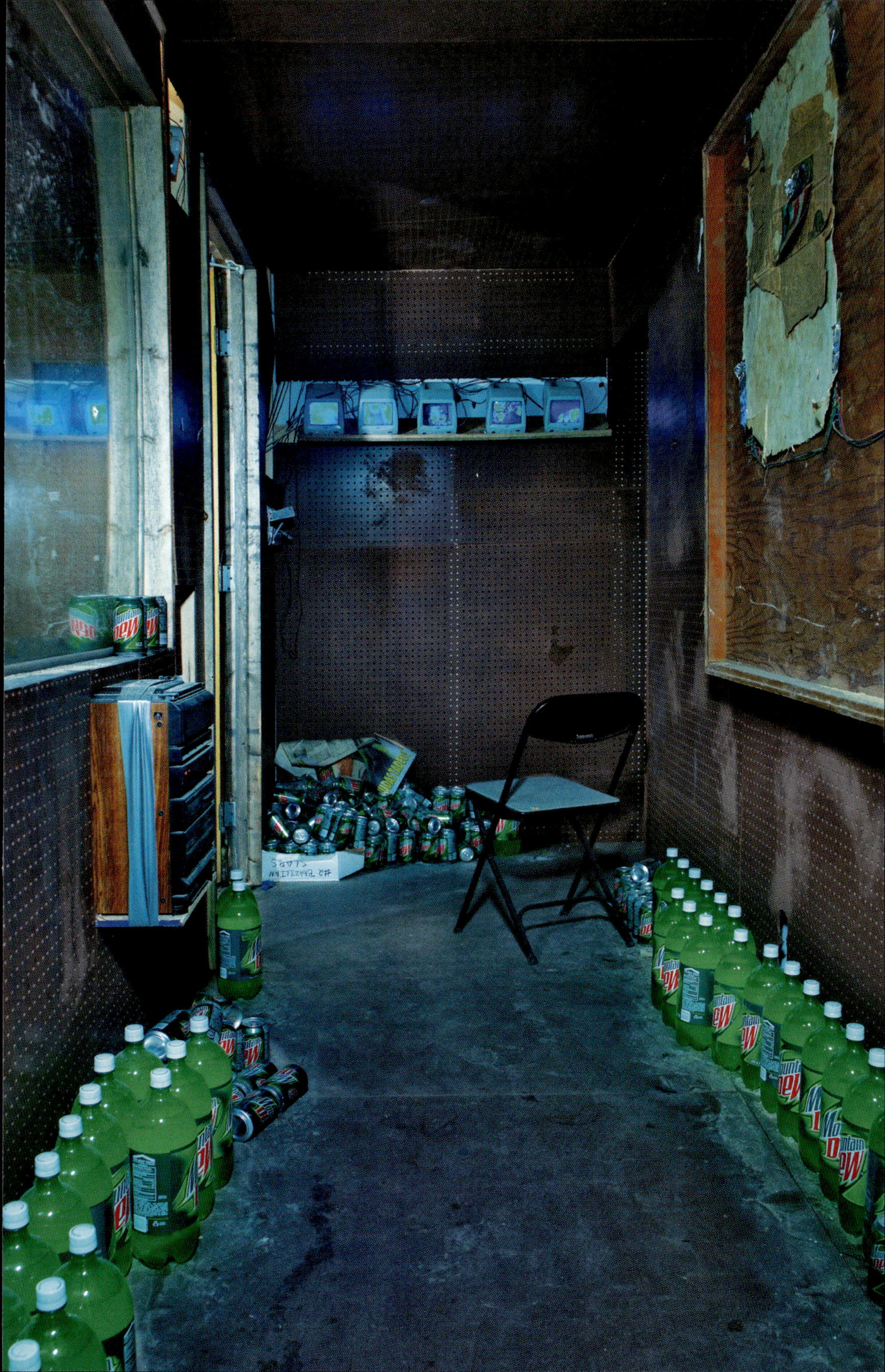

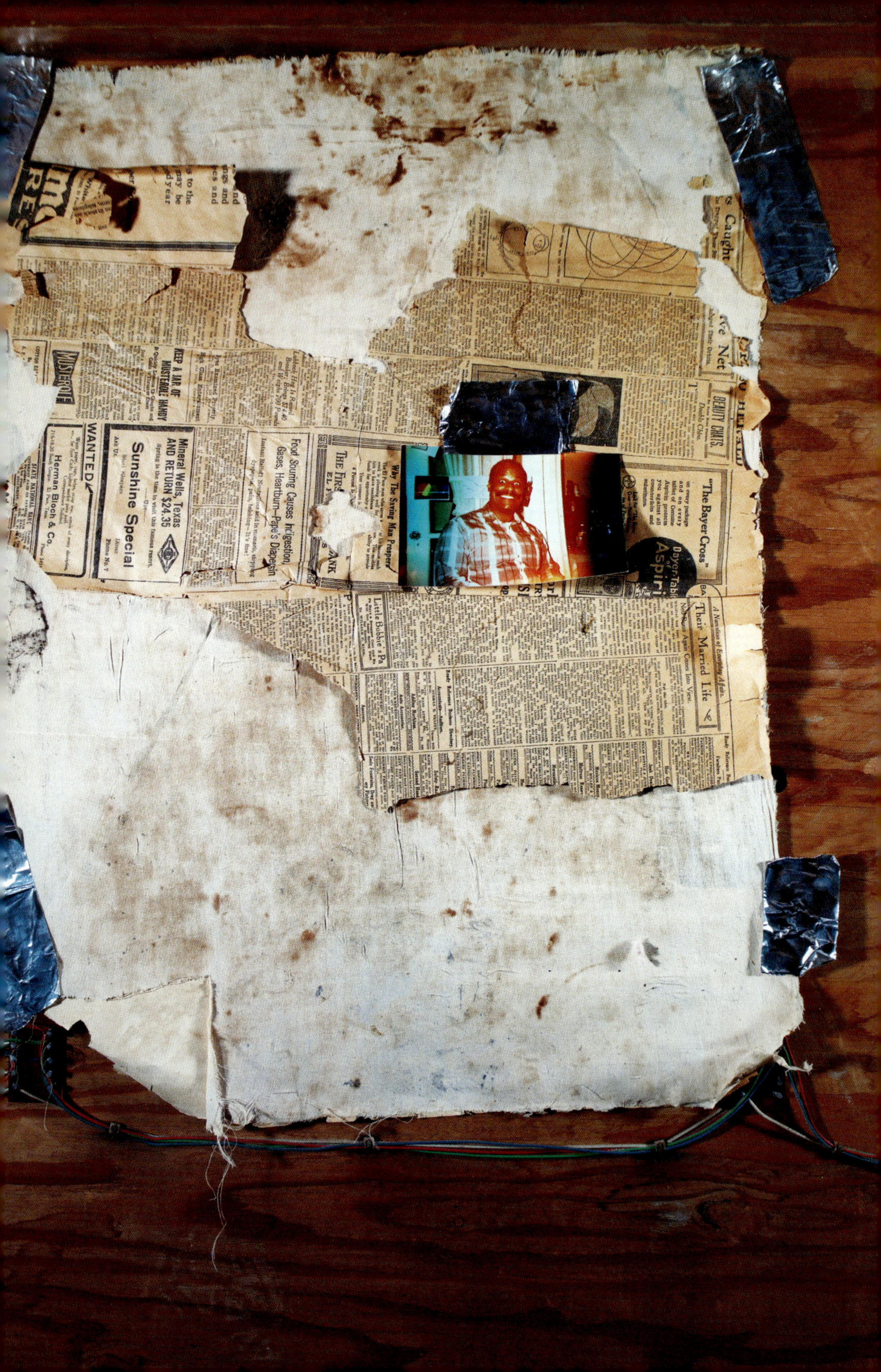

KEEP A JAR OF MUSTEROLE HANDY
WANTED
Herman Bloch & Co.
Mineral Wells, Texas
AND RETURN $24.35
Sunshine Special
Foot Souring Causes Indigestion, Gases, Heartburn—Pape's Diapepsin
Way The Saving Man Prospers
Little Bobbie Pa
"The Bayer Cross"
Their Married Life
BEAUTY CHATS

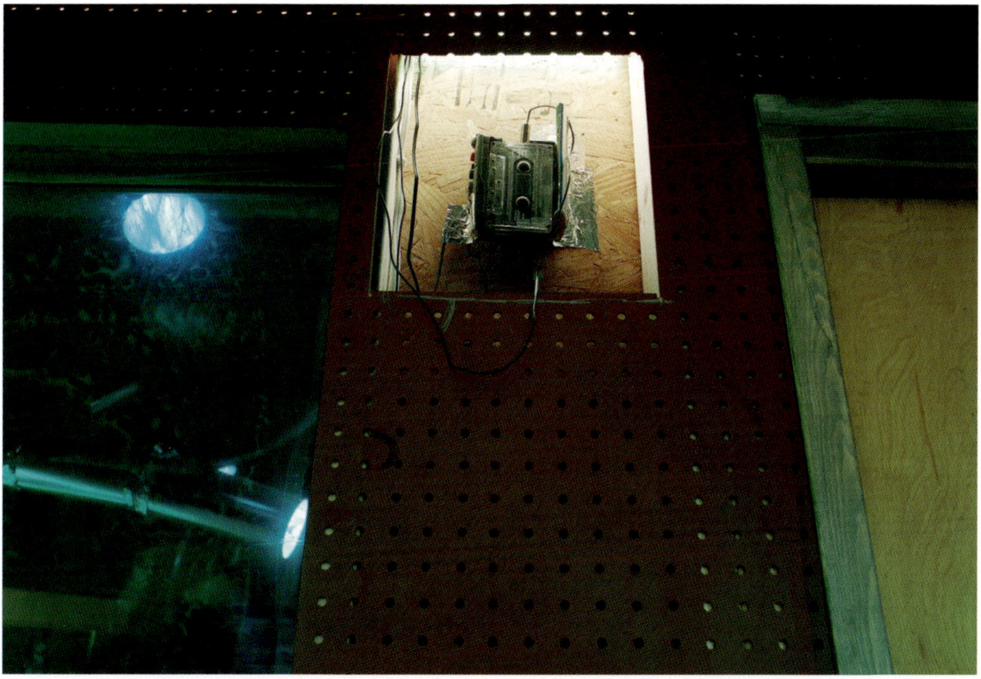

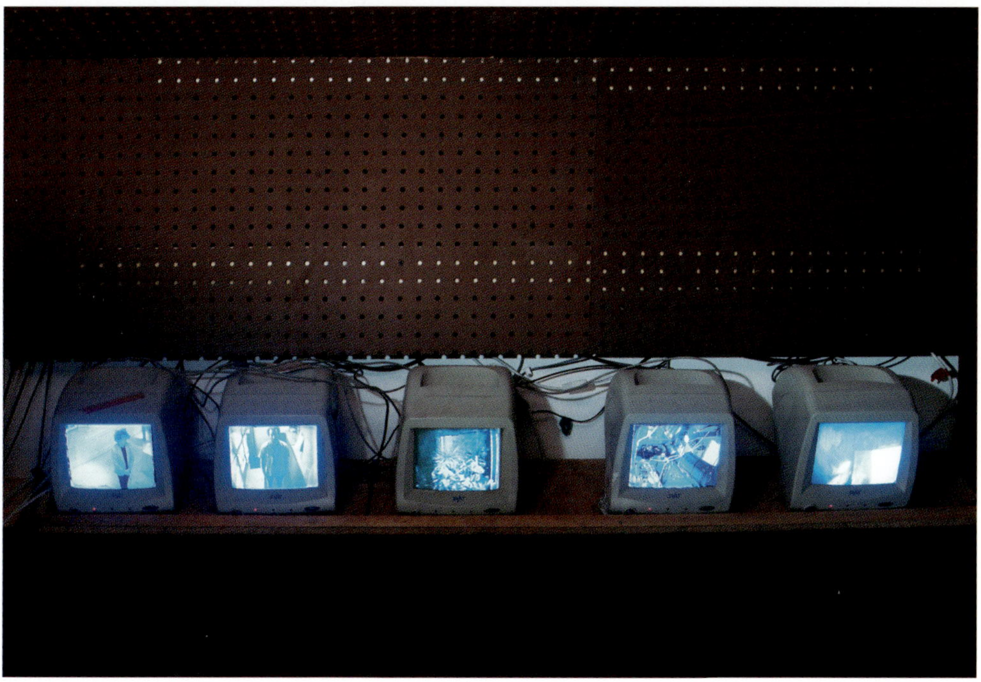

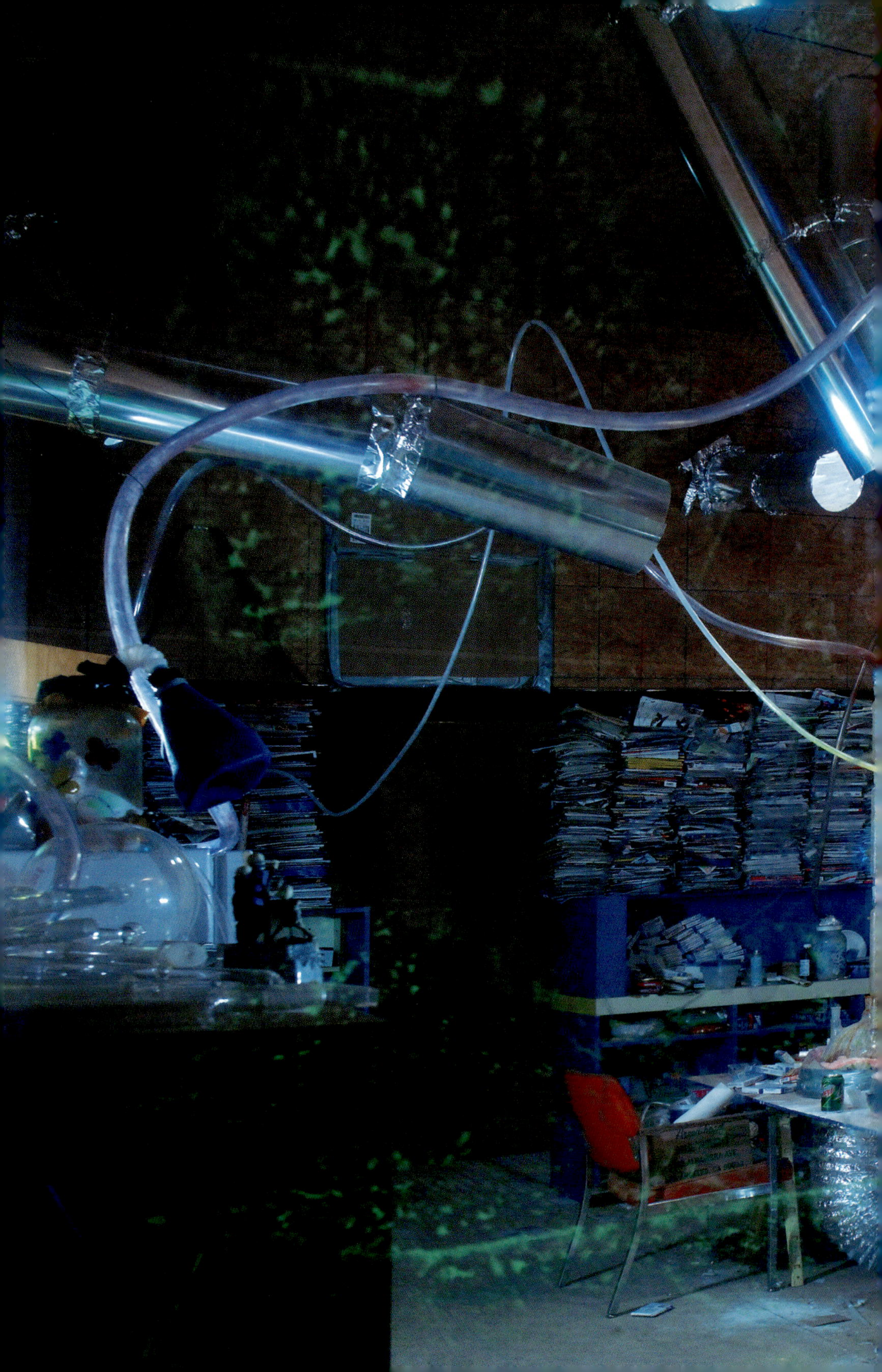

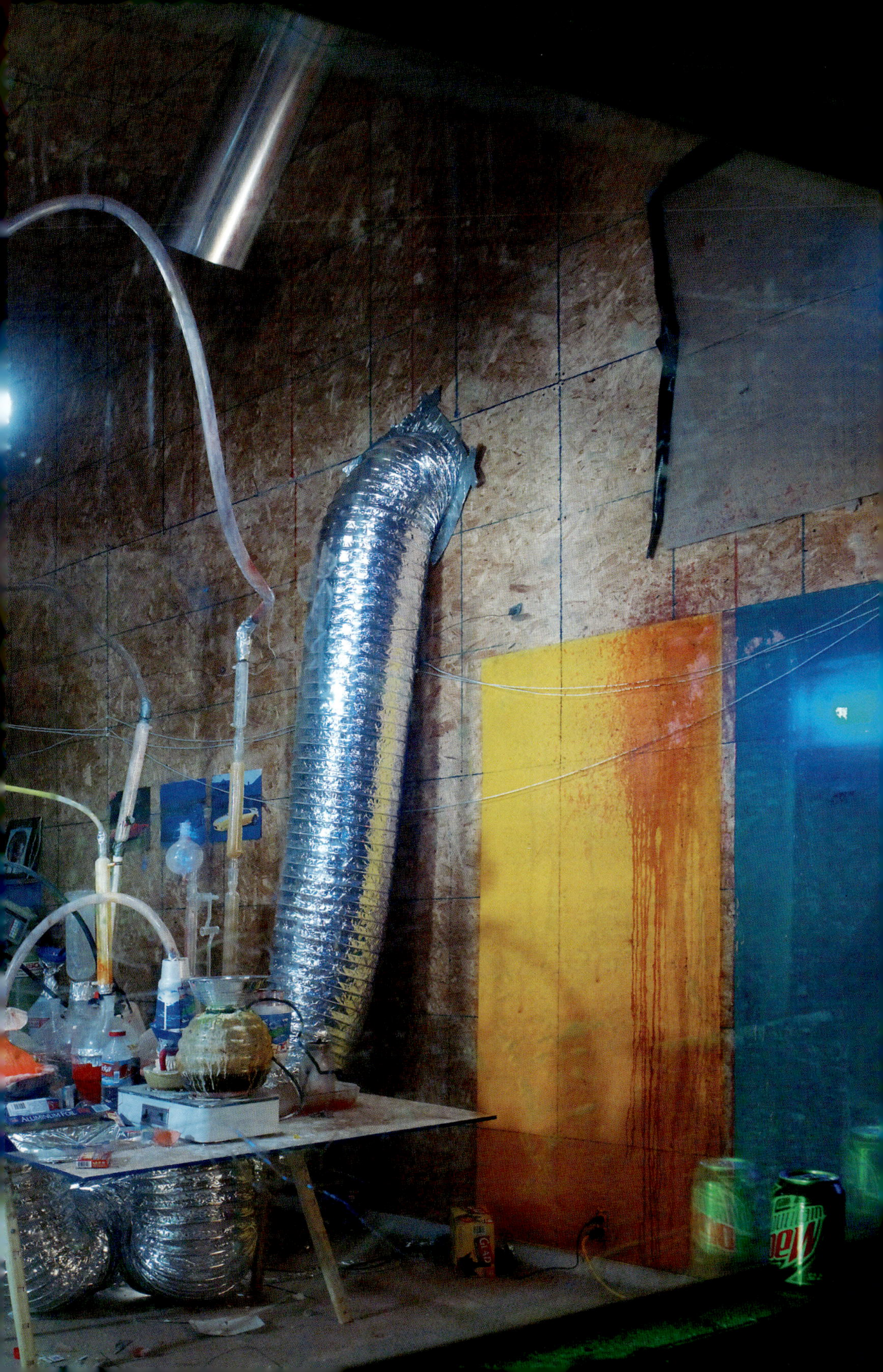
ALUMINUM FOIL
Glad

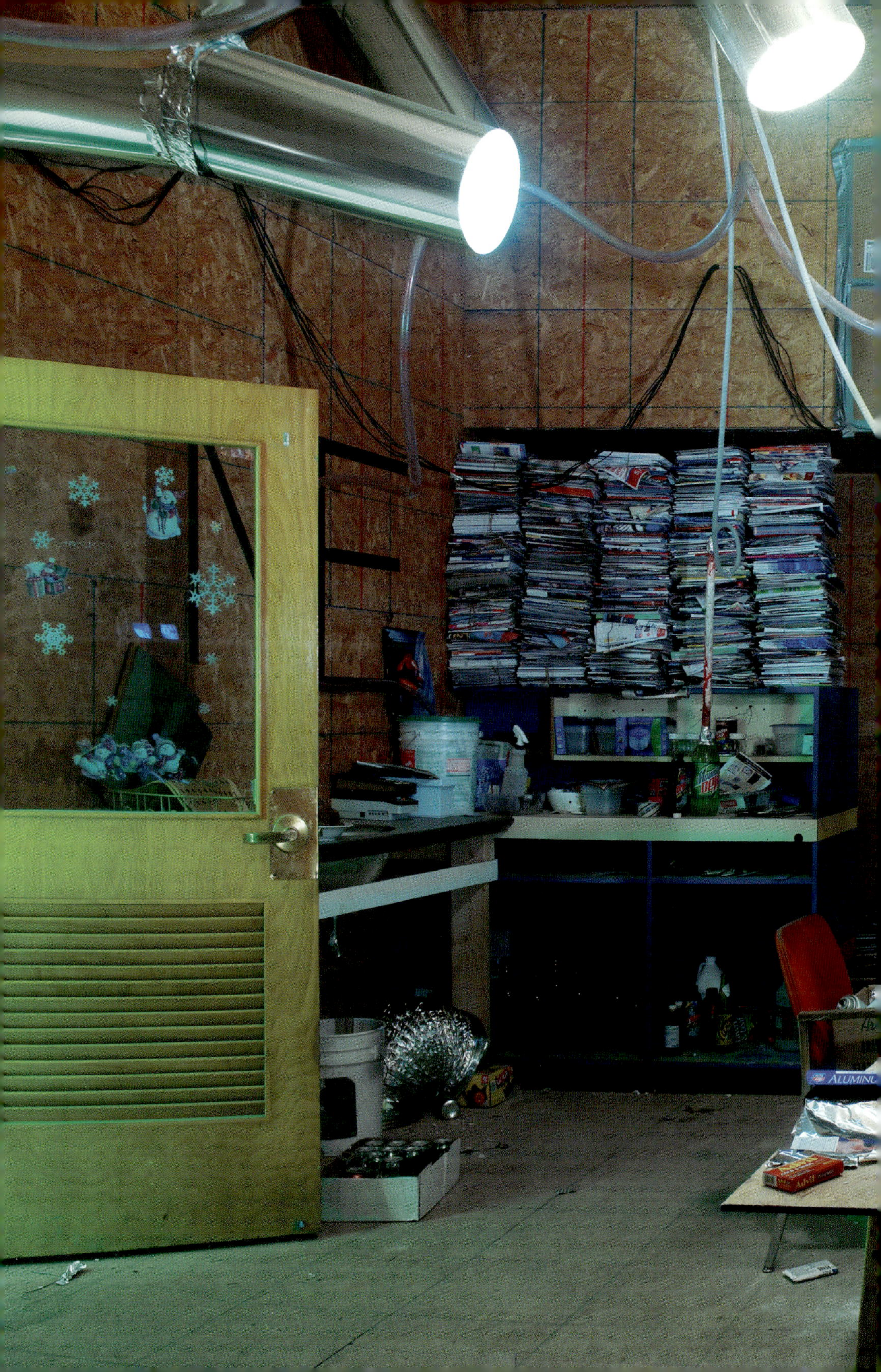
ALUMINU
Advil

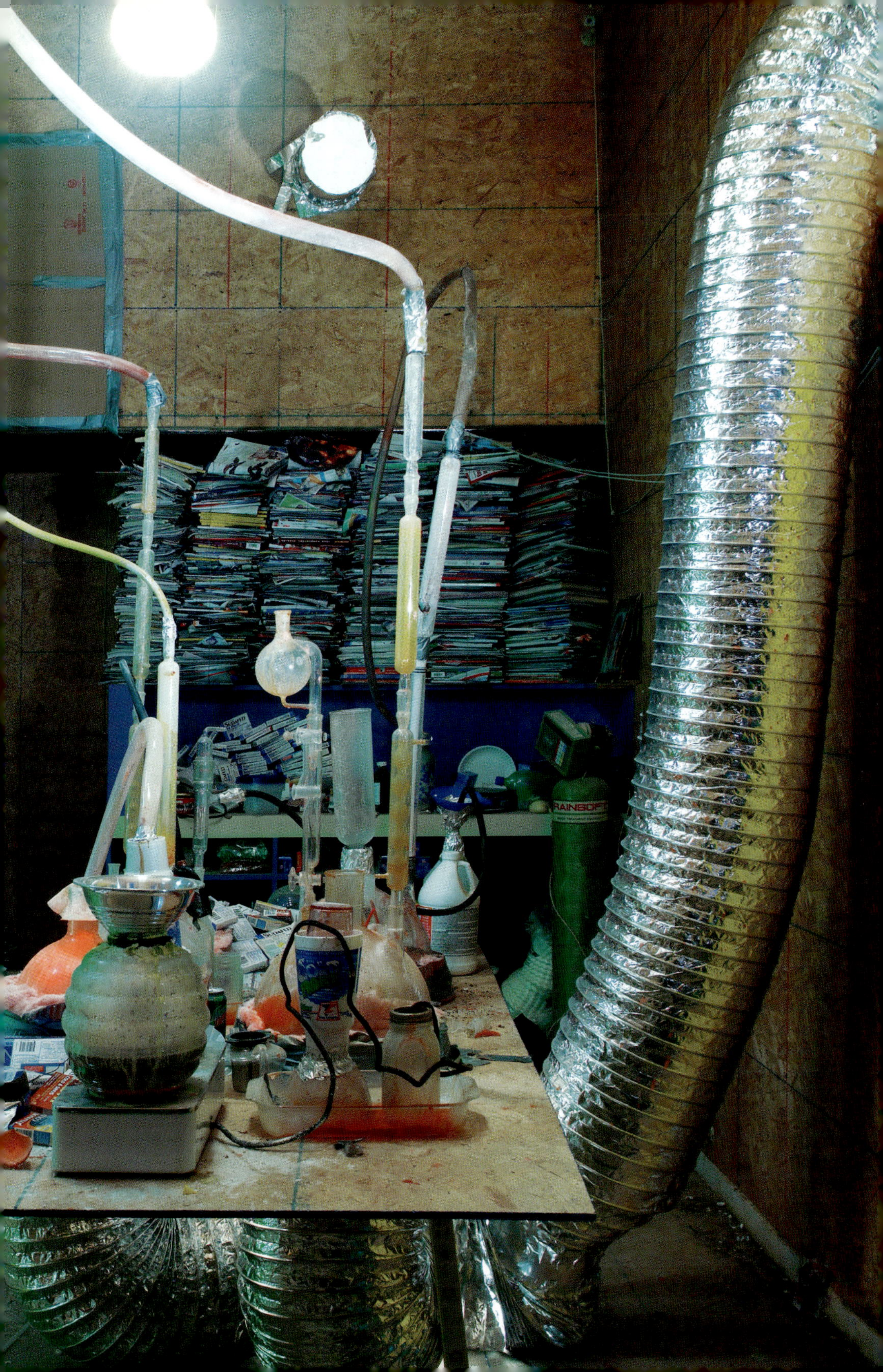
RAINSOFT

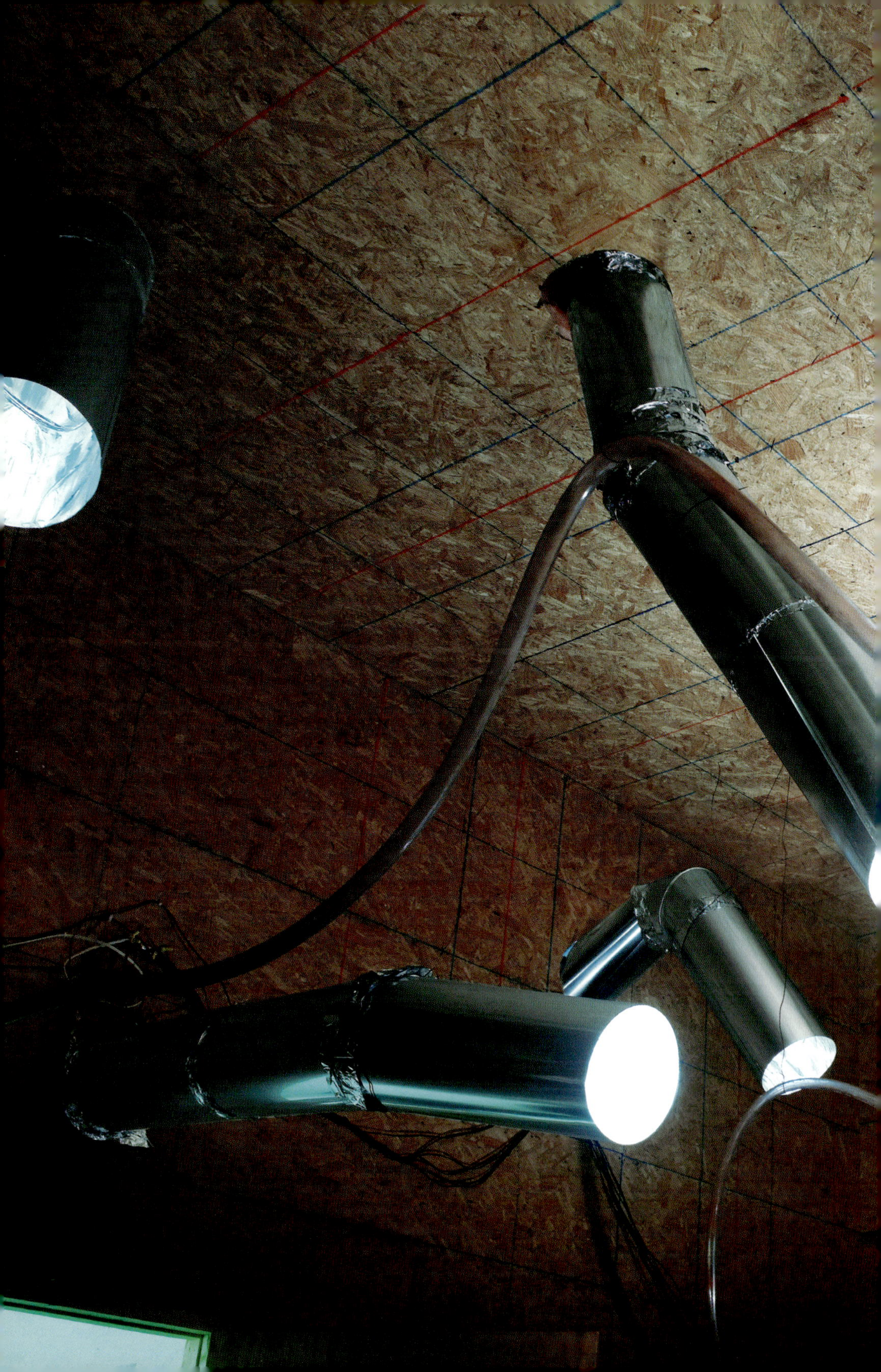

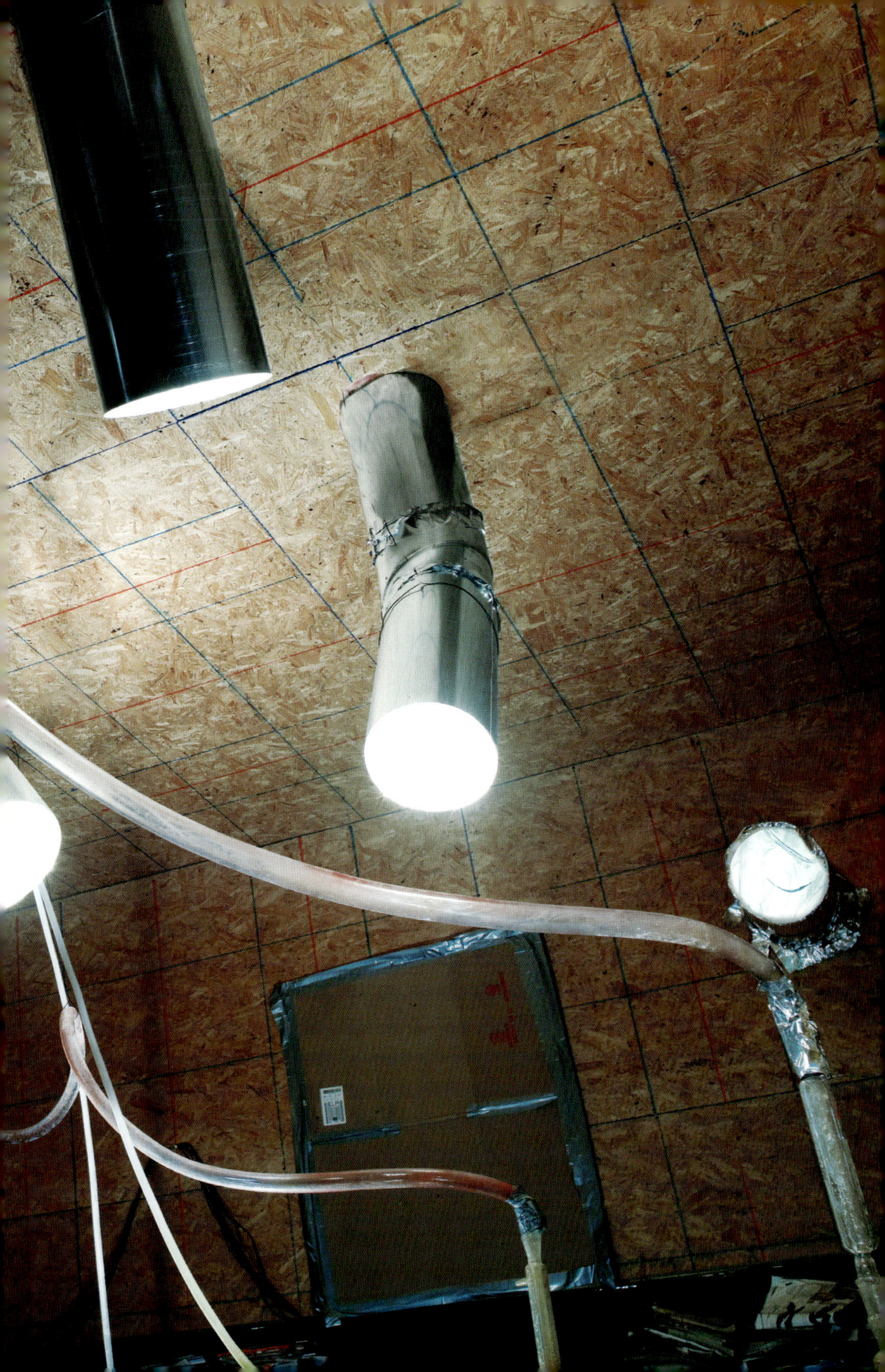

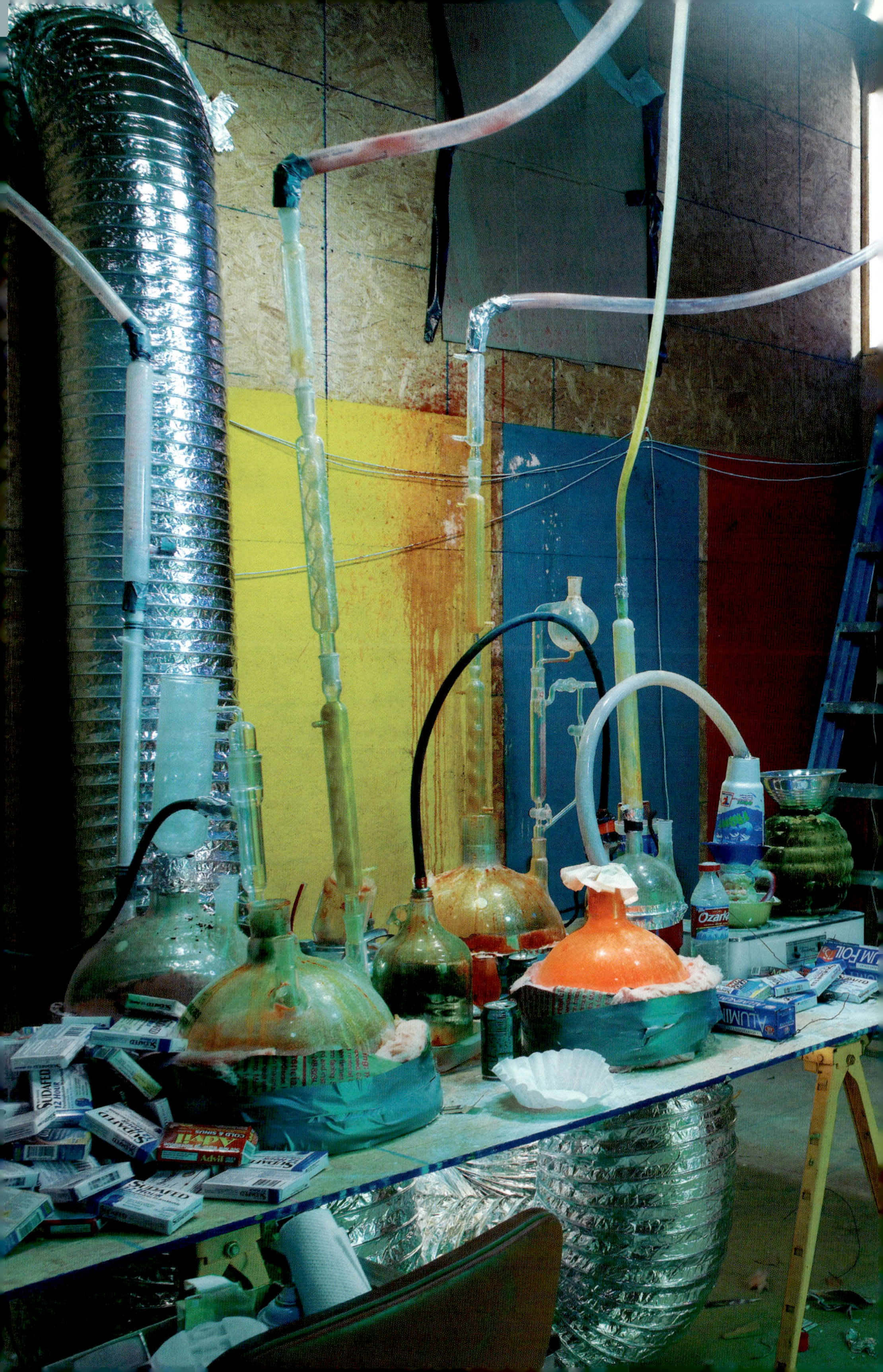
SUDAFED
12 Hour
Advil
COLD & SINUS
Ozarka

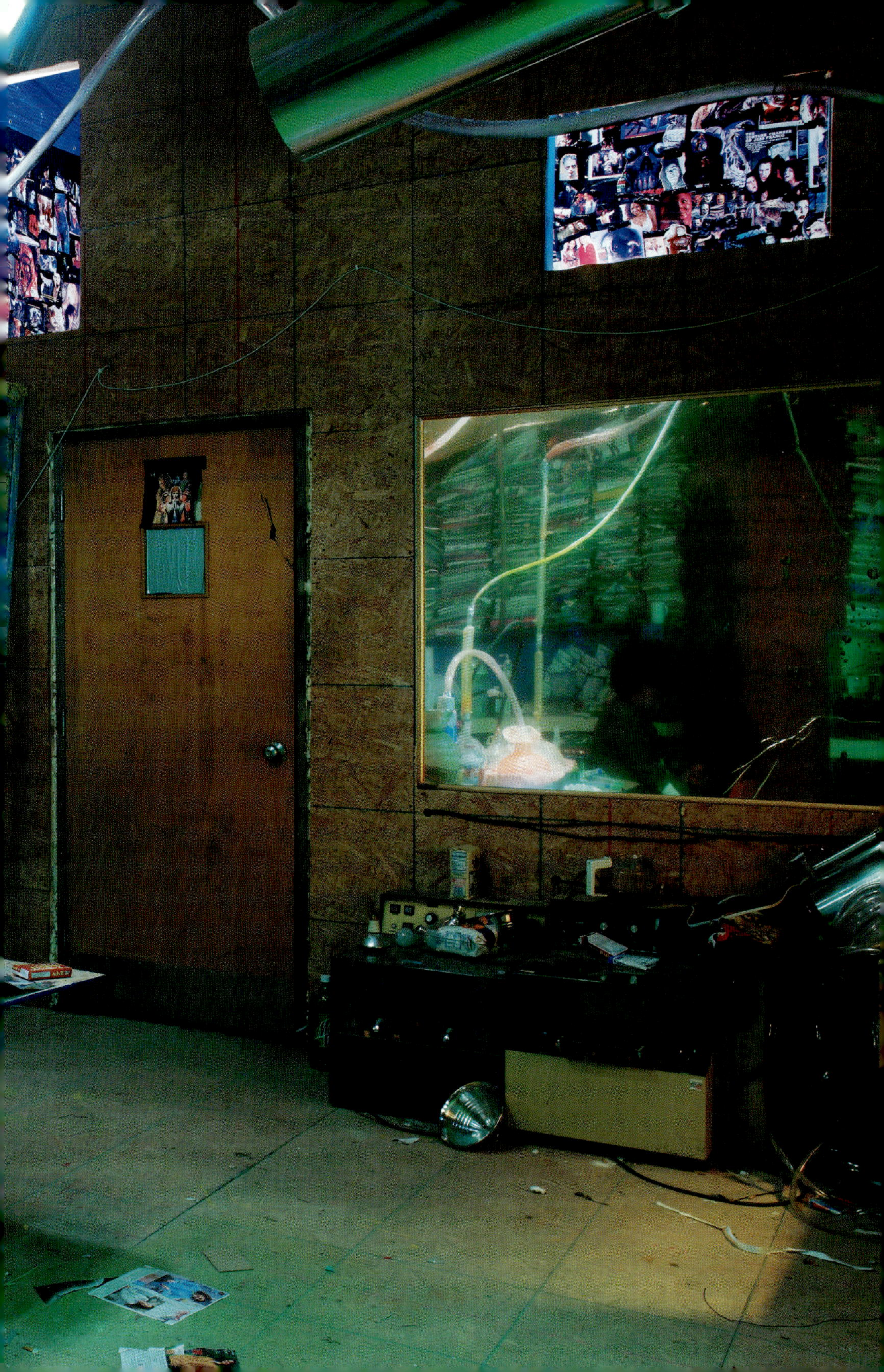

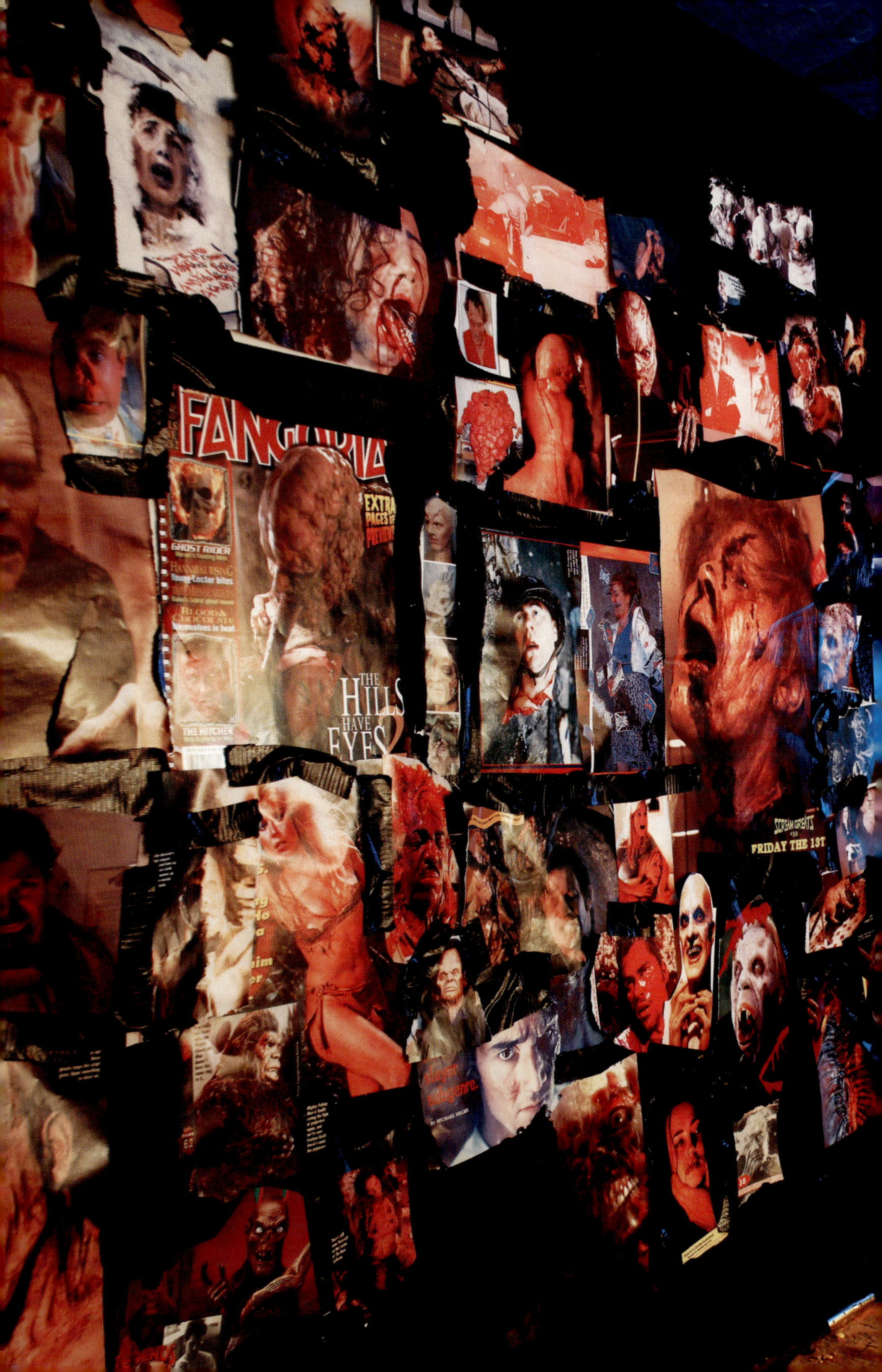
FANGORIA
EXTRA PAGES OF PREVIEWS
GHOST RIDER
Young Lecter bites
BLOOD & CHOCOLATE
THE HITCHER
THE HILLS HAVE EYES 2
SCREAM GREATS
FRIDAY THE 13TH

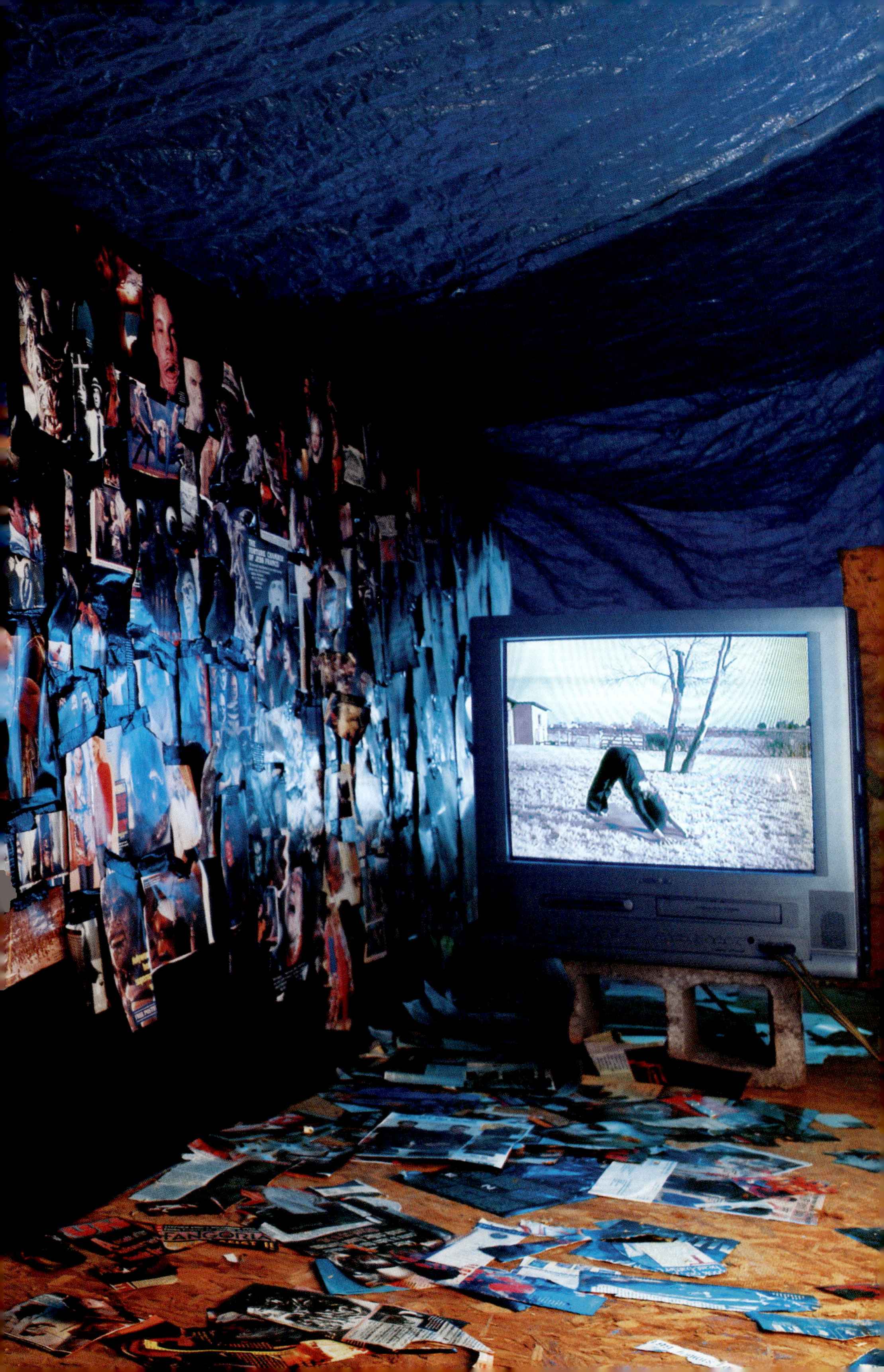

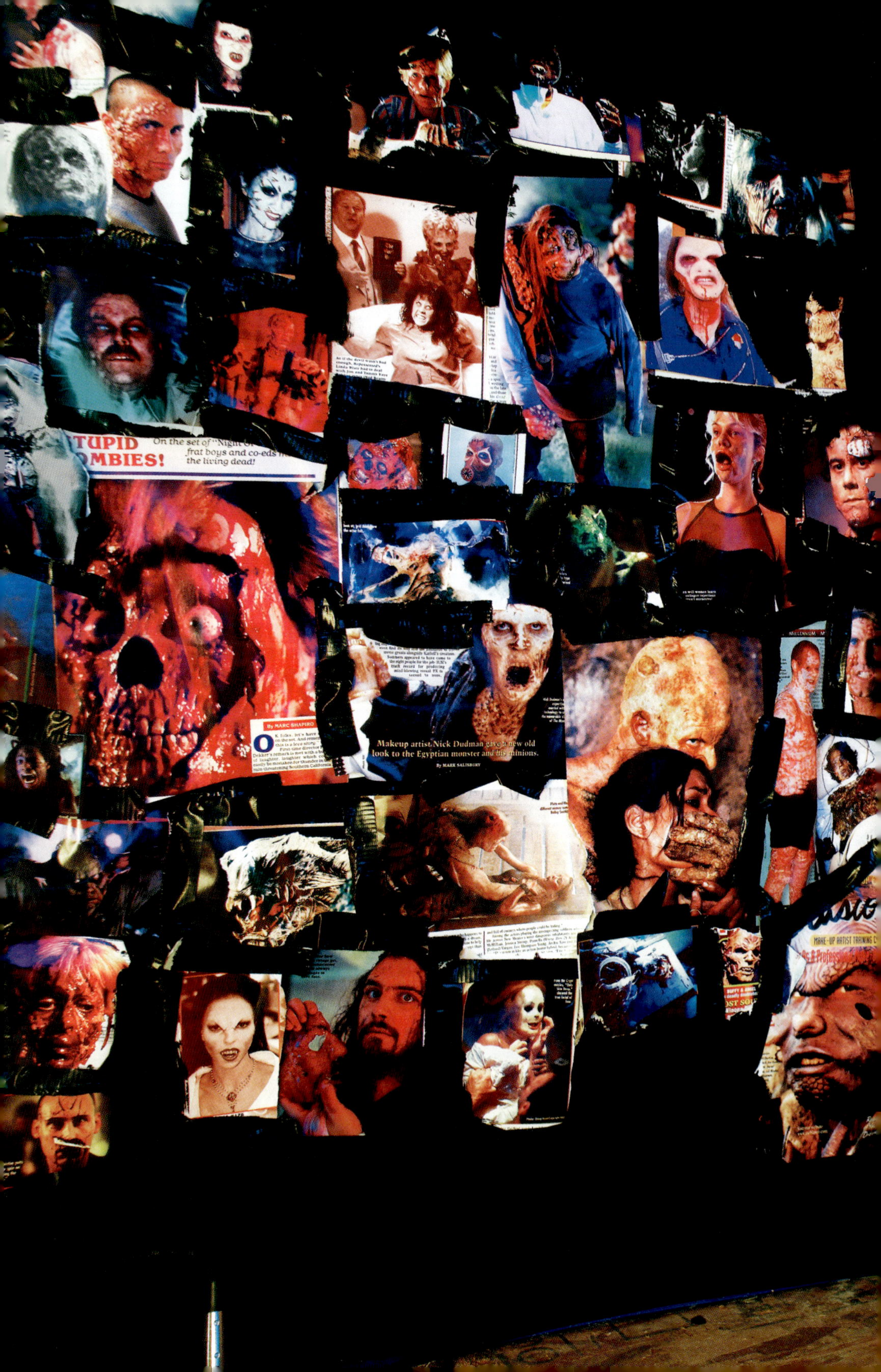
TUPID
OMBIES!
the living dead!
By MARC SHAPIRO
Makeup artist Nick Dudman gave a new old look to the Egyptian monster and his minions.
By MARK SALISBURY

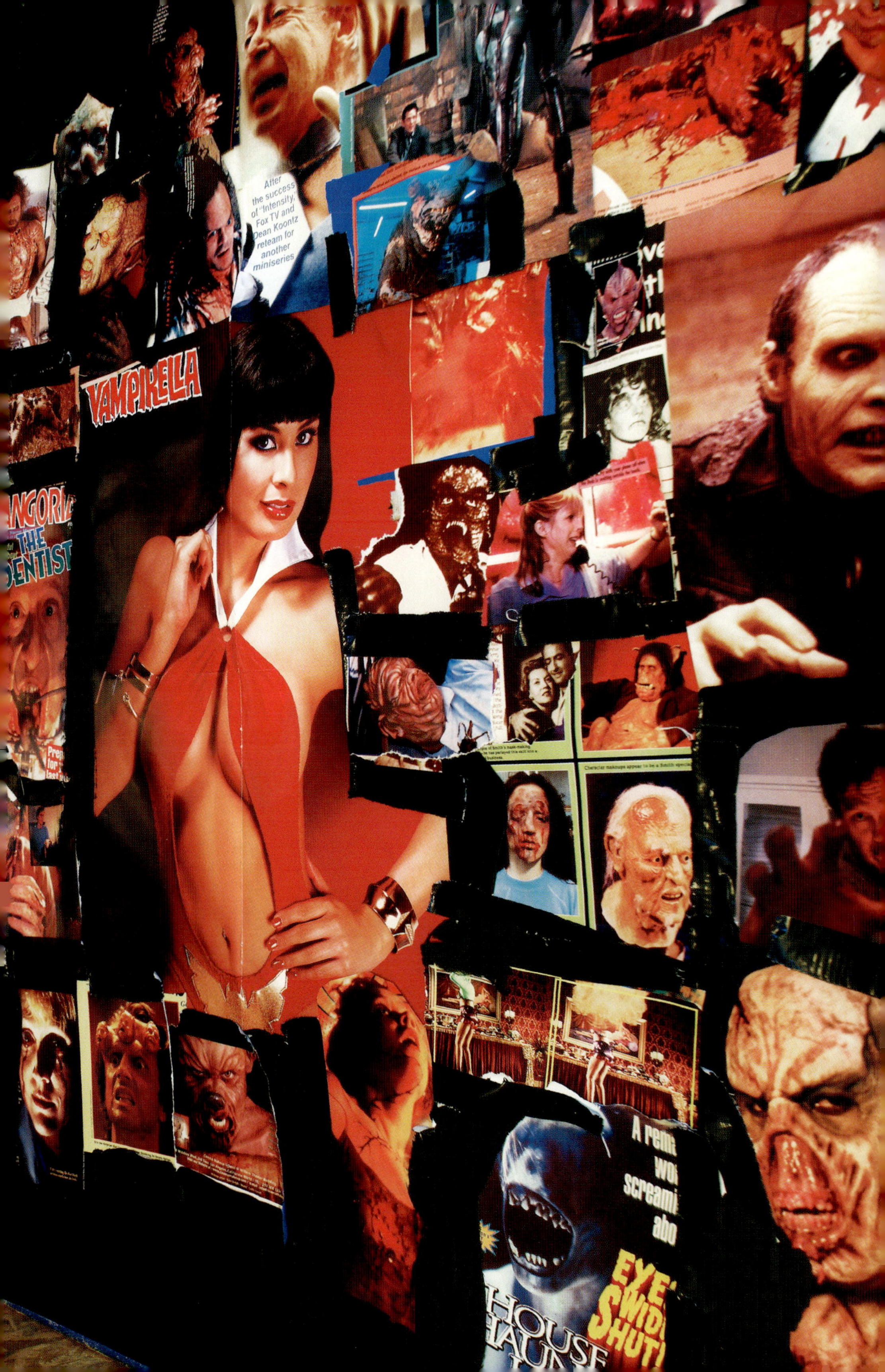
After the success of "Intensity," Fox TV and Dean Koontz reteam for another miniseries
VAMPIRELLA
THE DENTIST

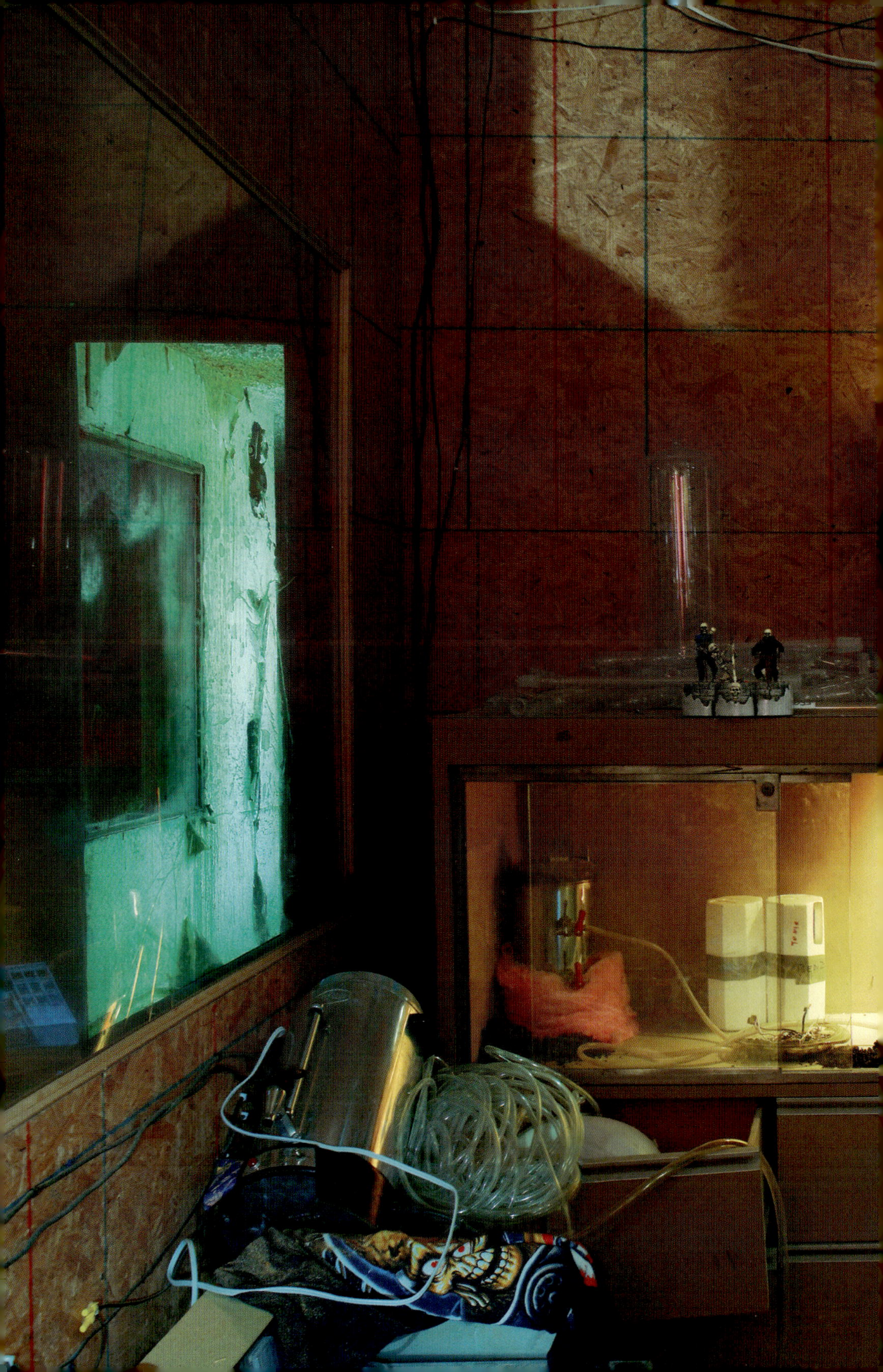

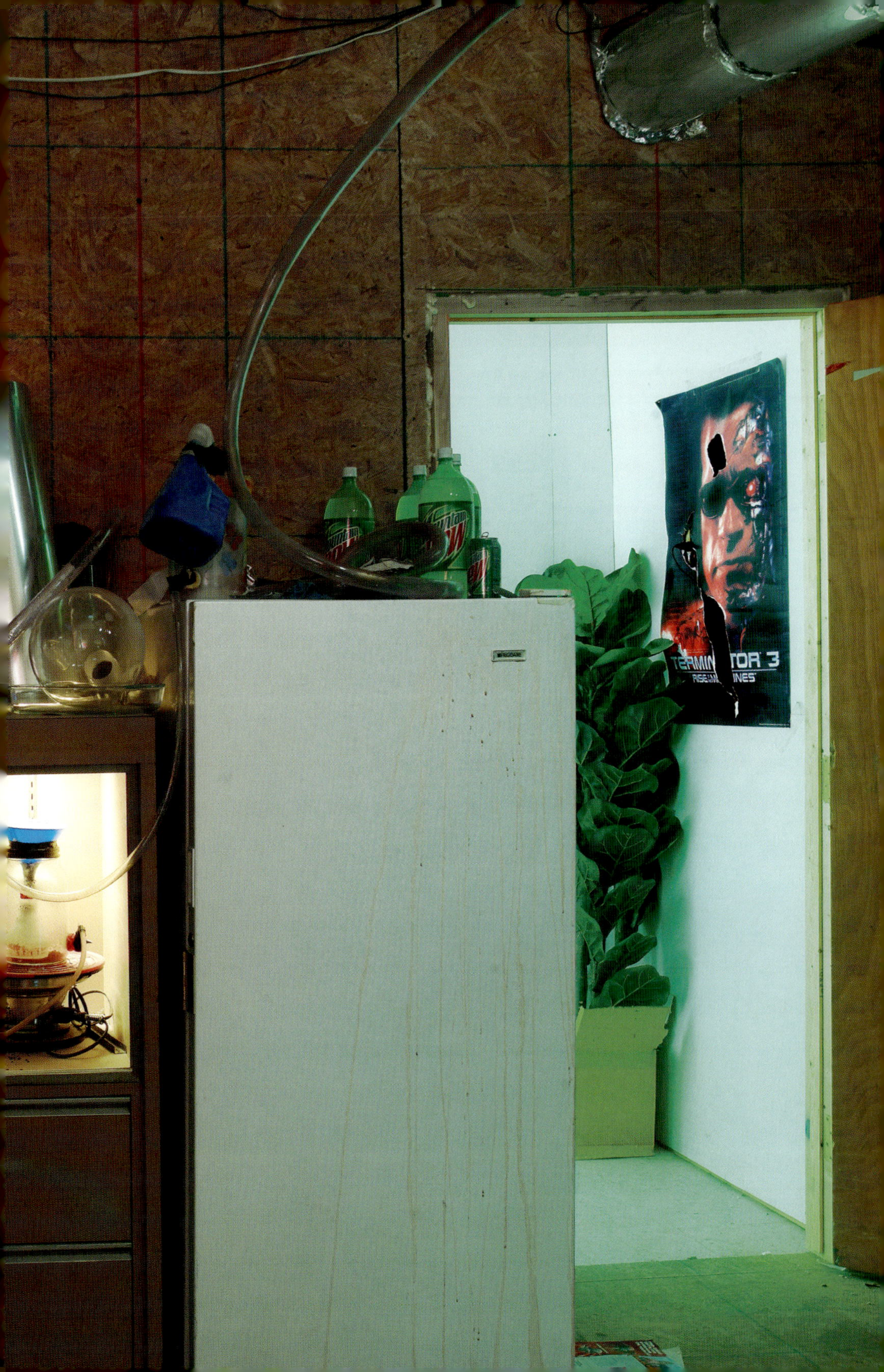
TERMIN TOR 3

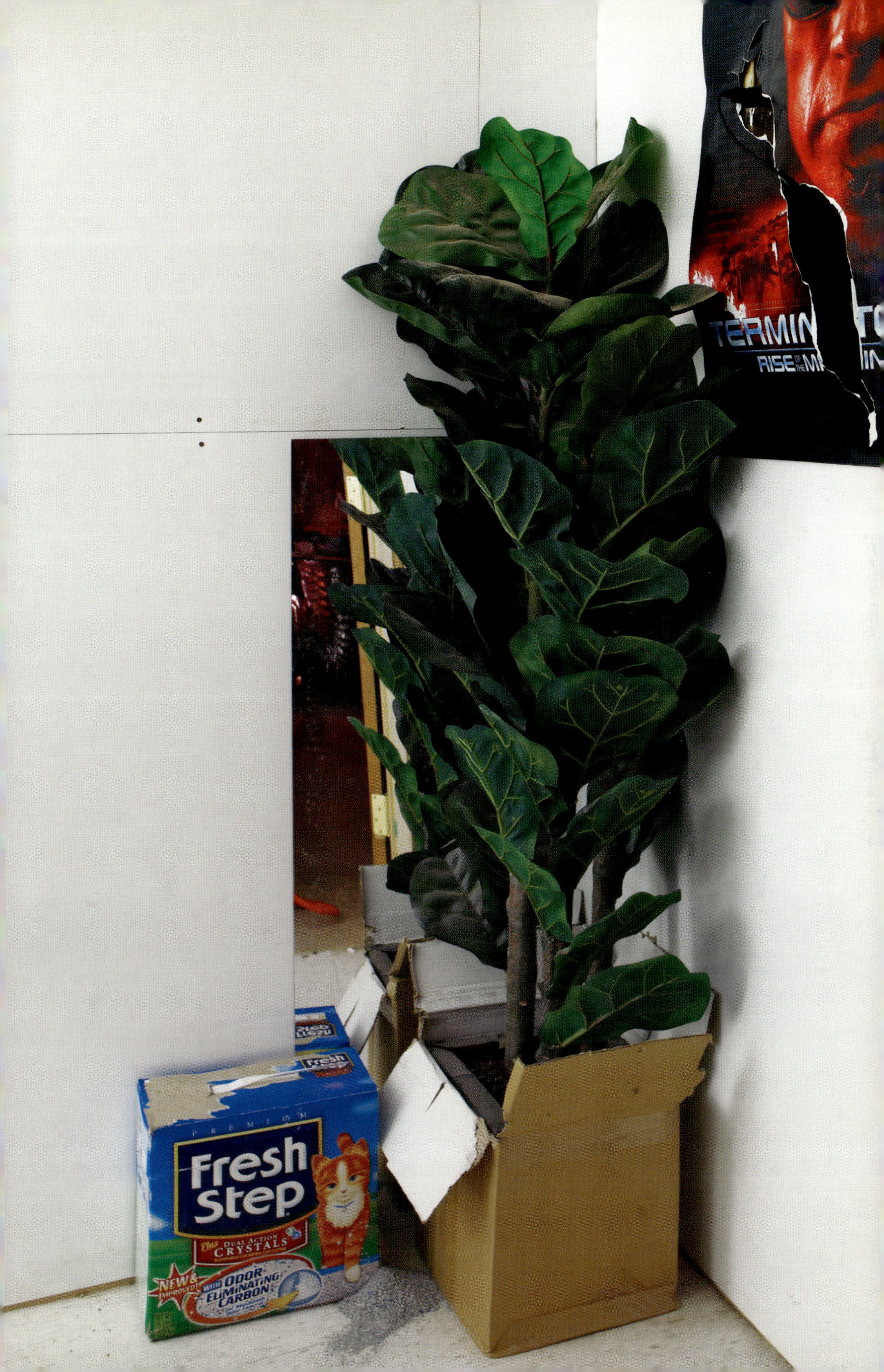
PREMIUM
Fresh Step
DUAL ACTION
CRYSTALS
NEW & IMPROVED
ODOR ELIMINATING CARBON
TERMIN
RISE

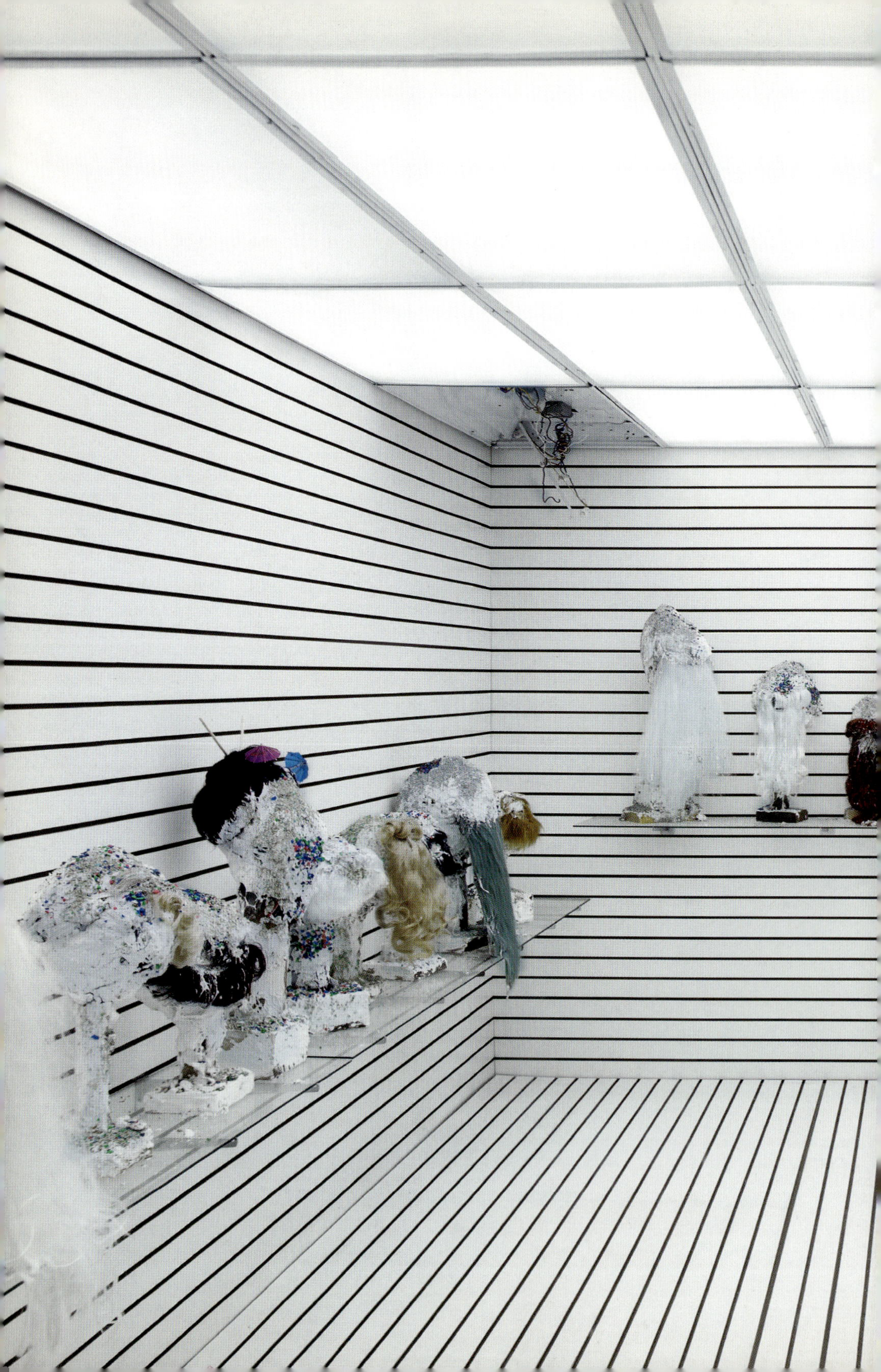

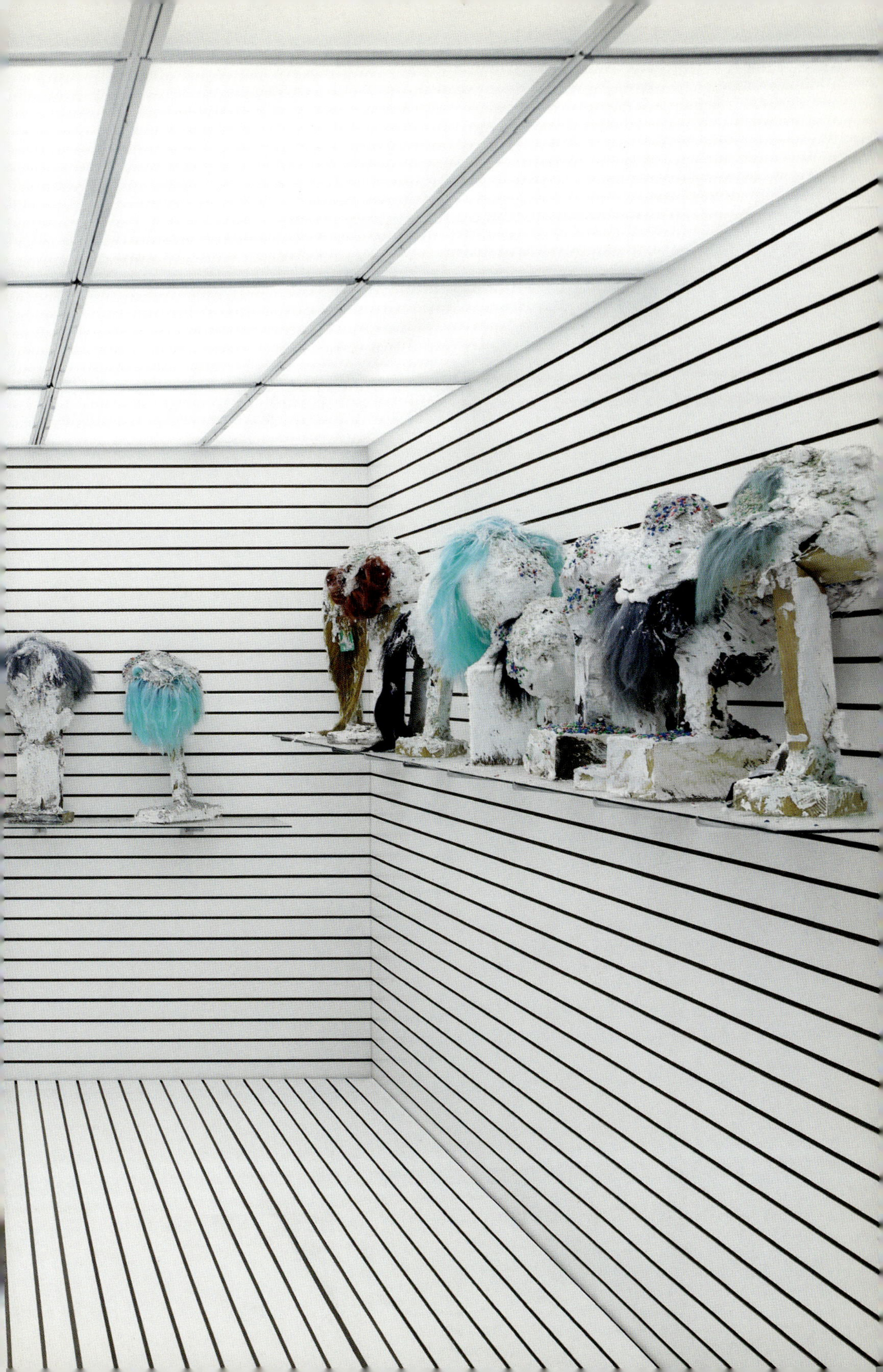

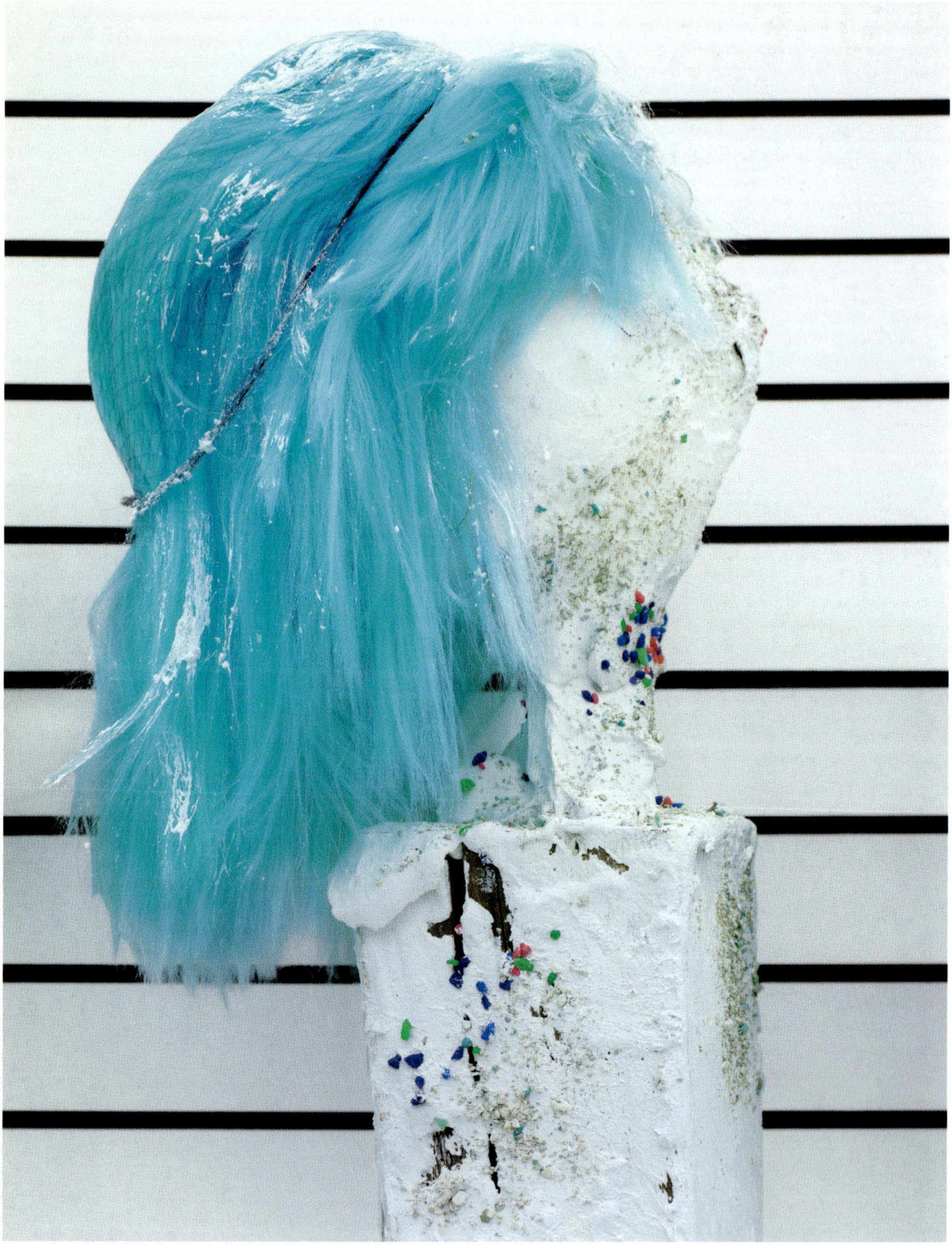

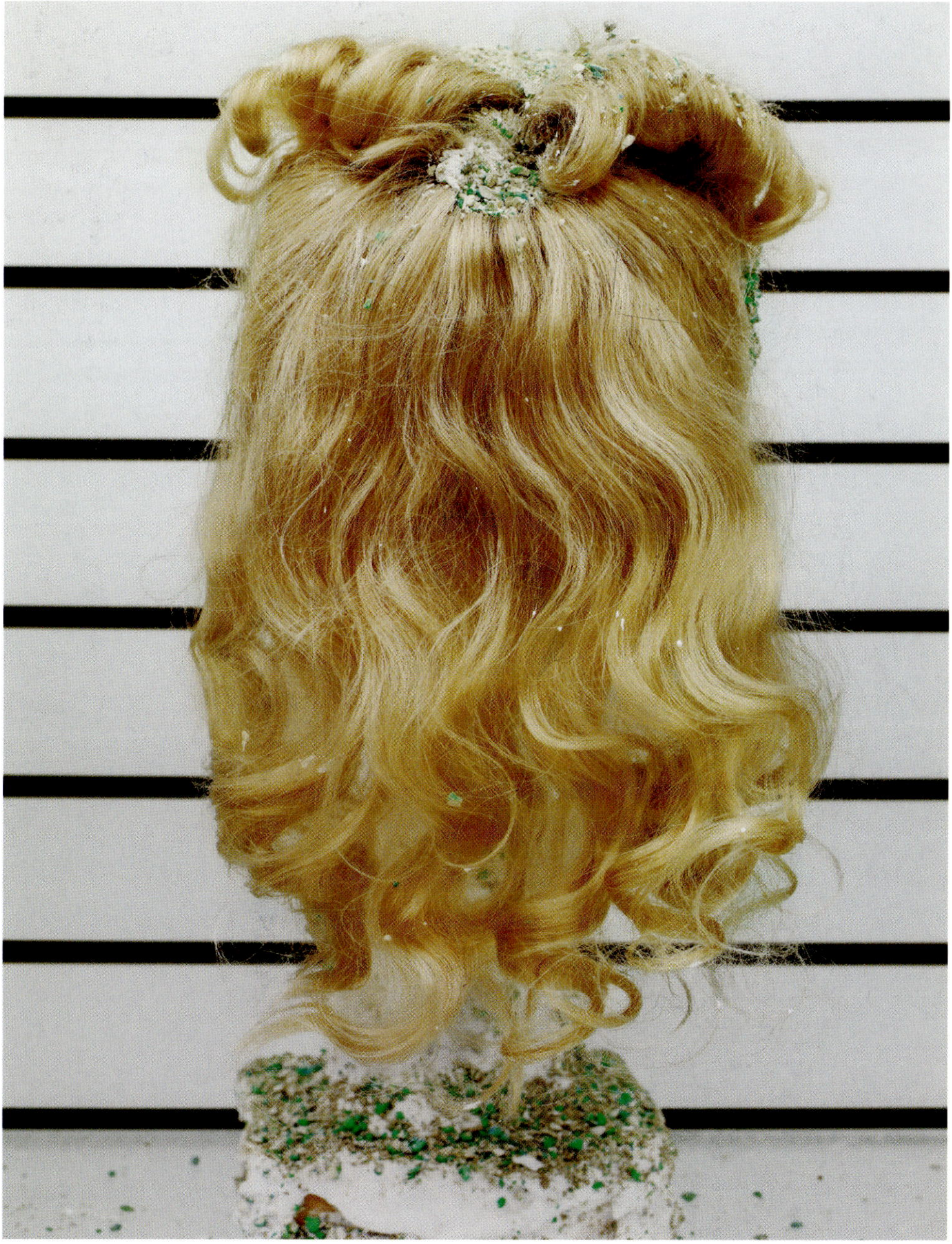

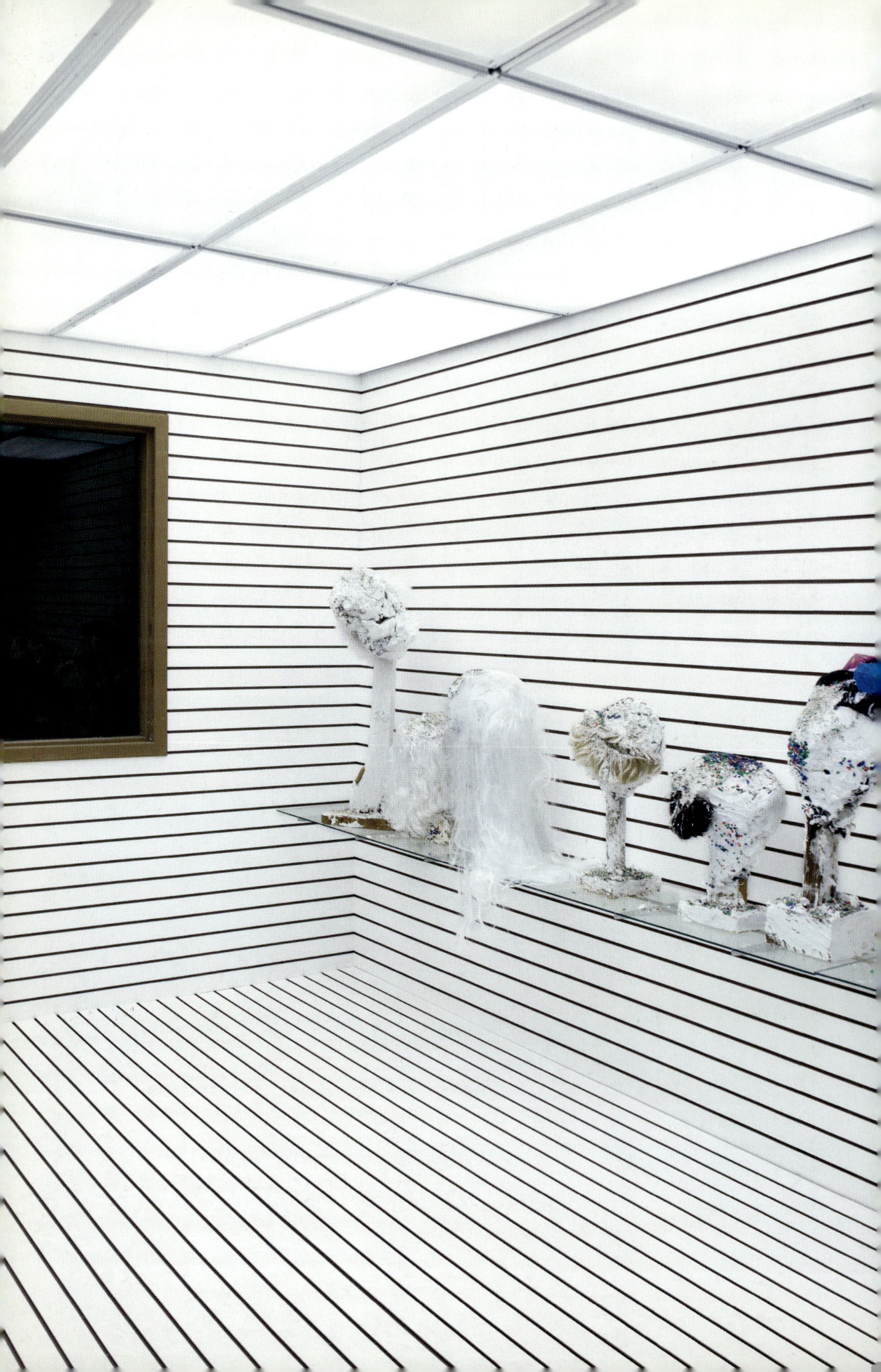

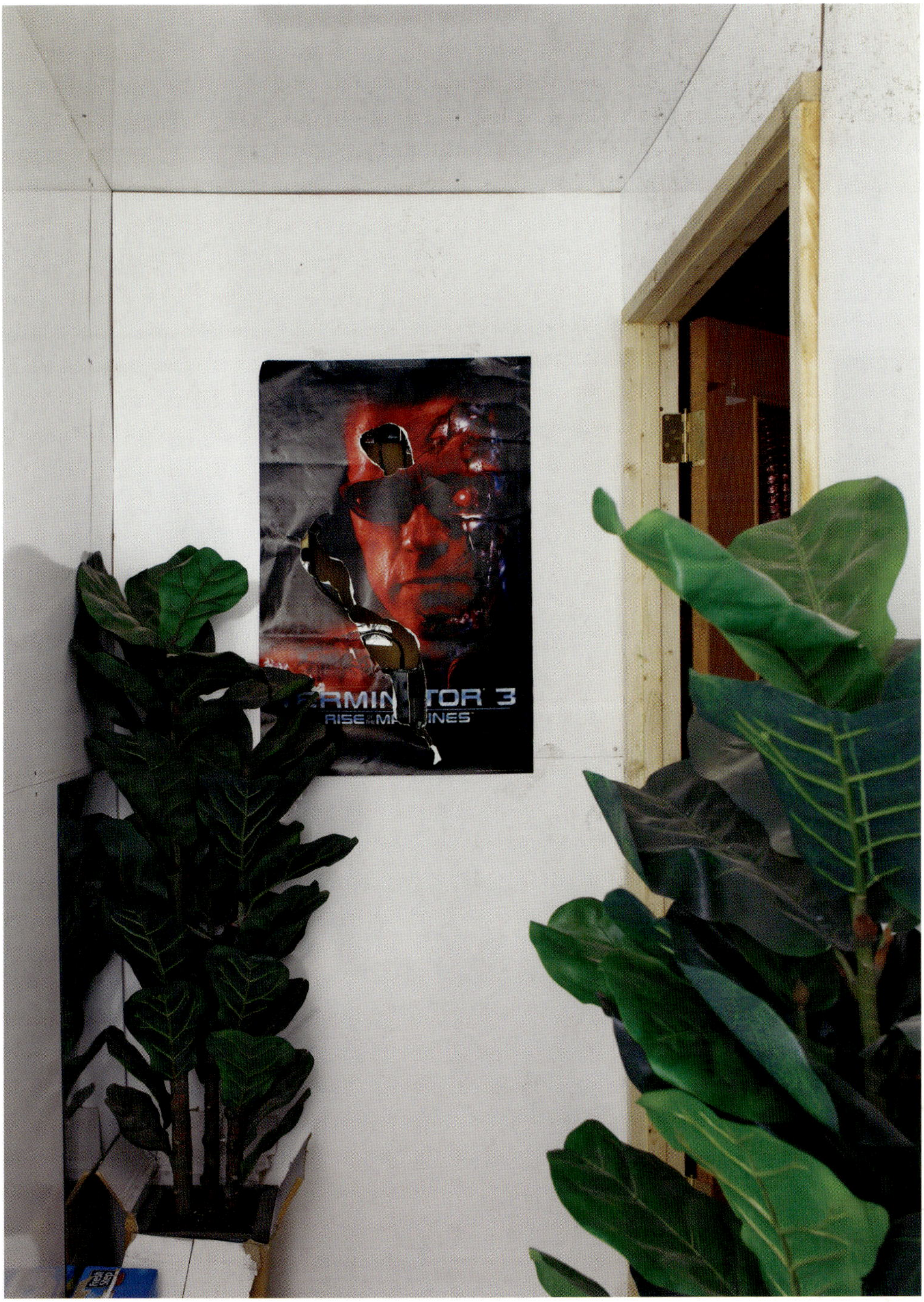

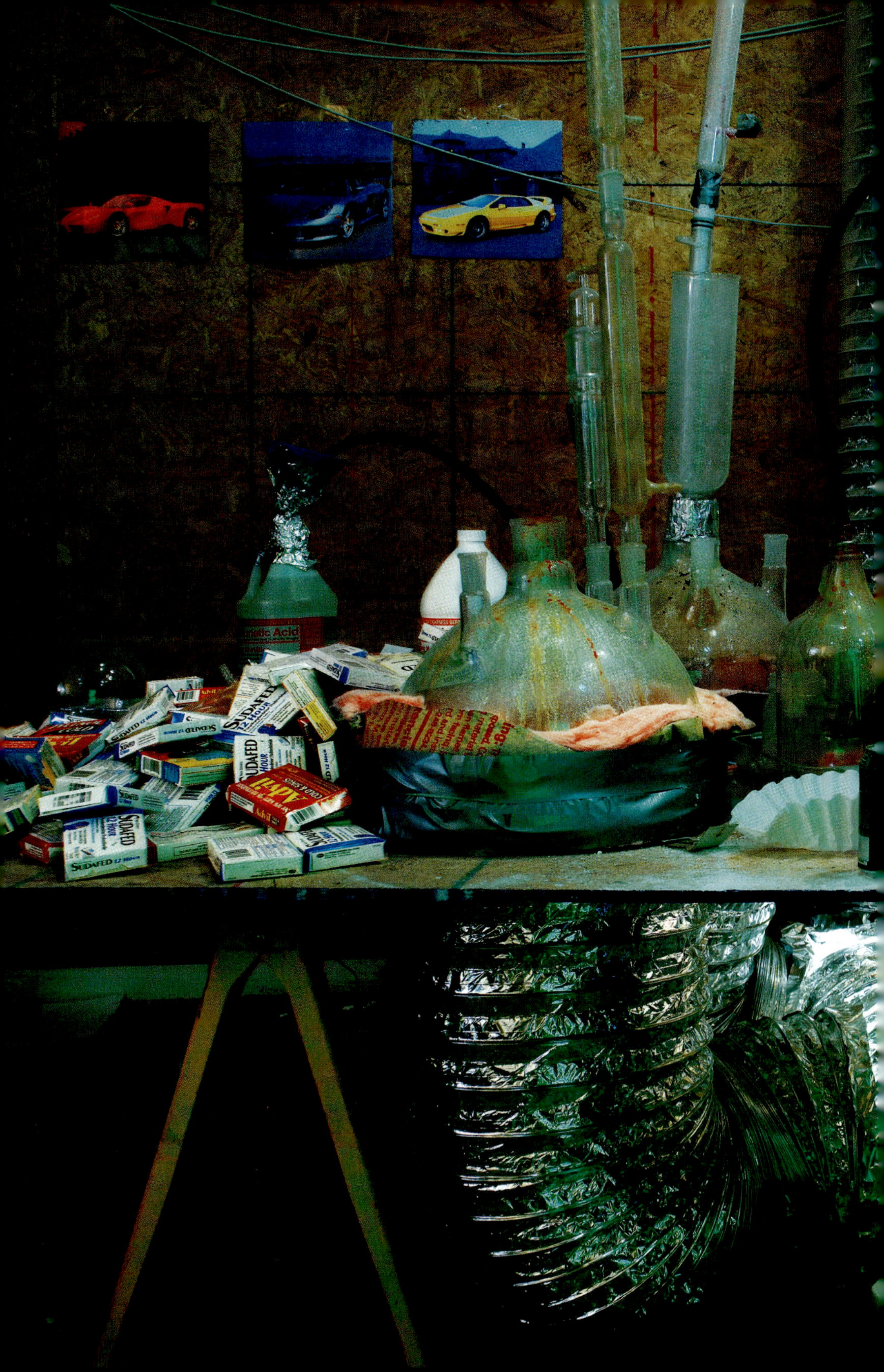
SUDAFED
12 Hour
Advil
COLD & SINUS

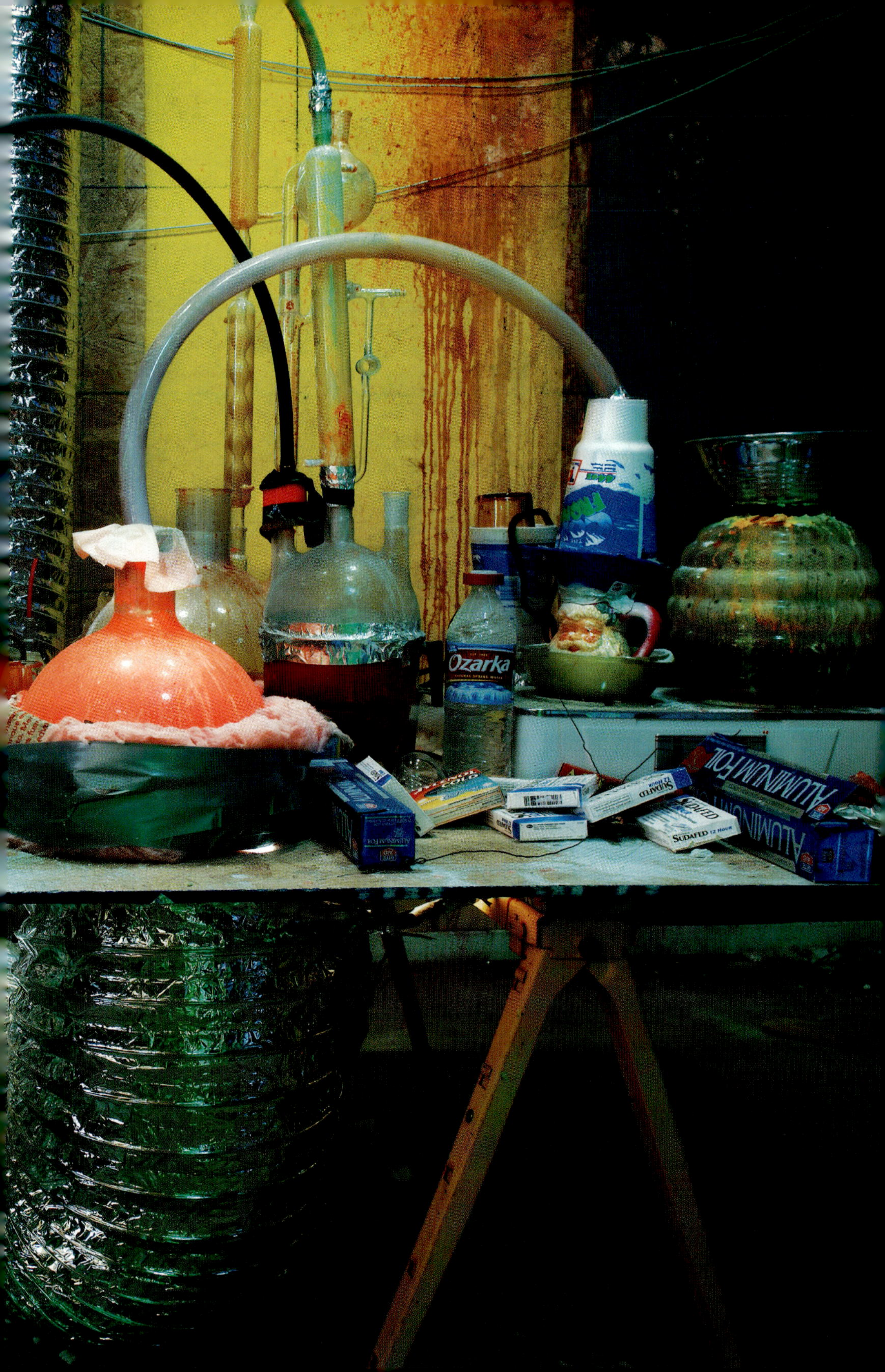
Ozarka
ALUMINUM FOIL
SUDAFED 12 Hour
SUDAFED 12 Hour

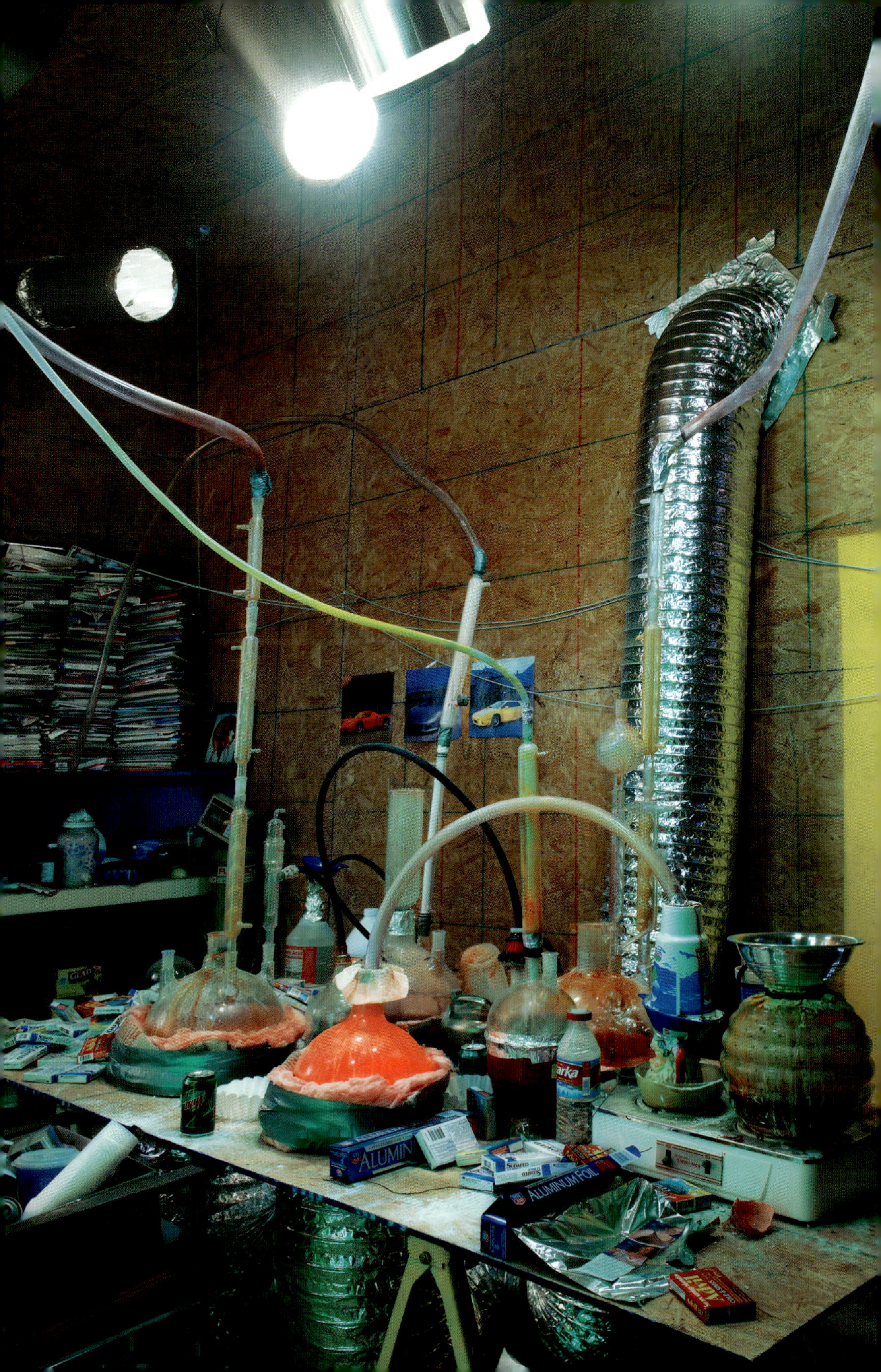
ALUMINUM FOIL

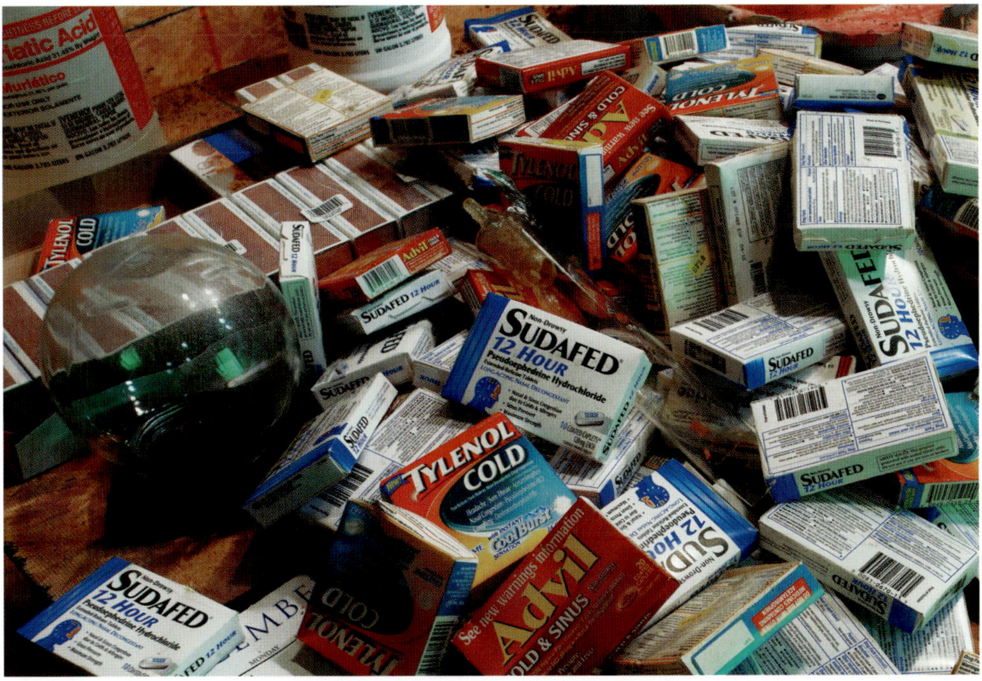
Muriatic Acid
Non-Drowsy
SUDAFED
12 HOUR
Pseudoephedrine Hydrochloride
TYLENOL
COLD
Cool Burst
See new warnings information
Advil
COLD & SINUS

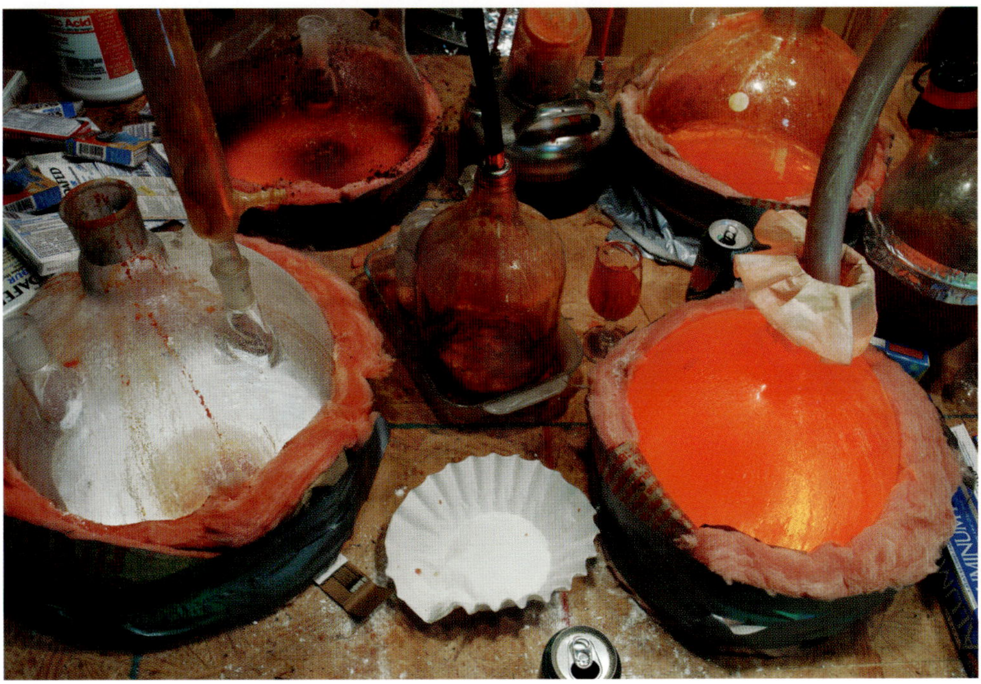

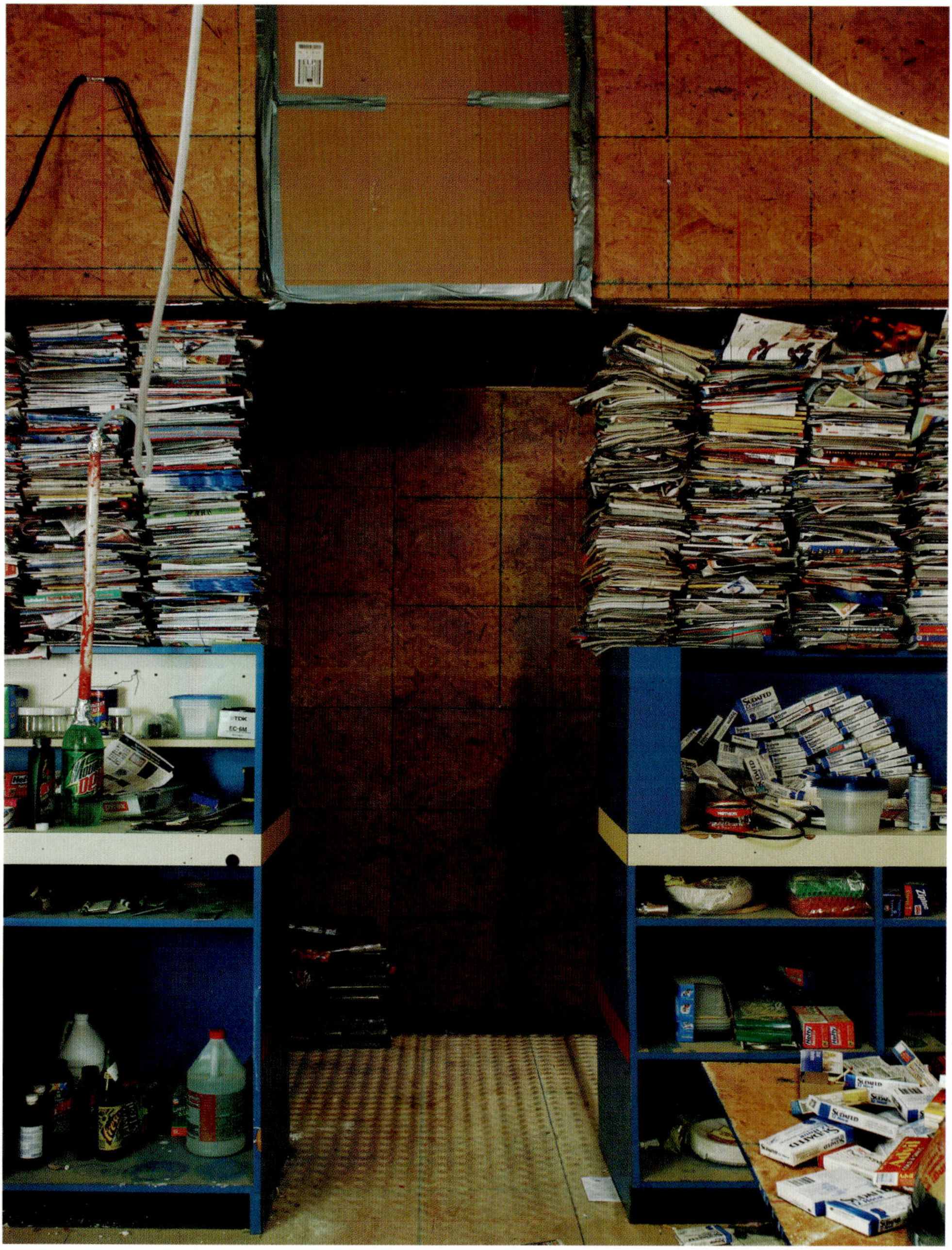
TDK
EC-6M

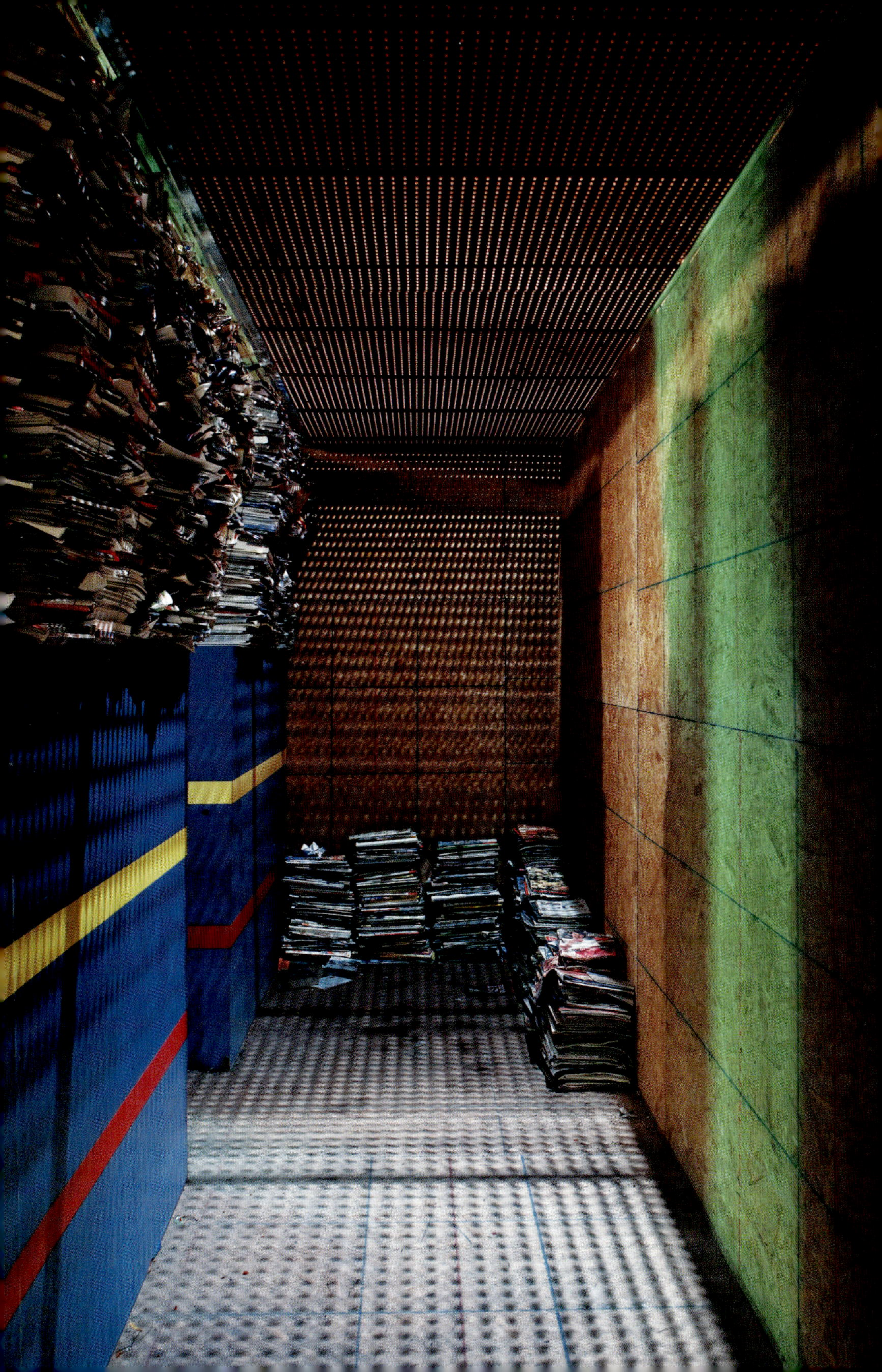

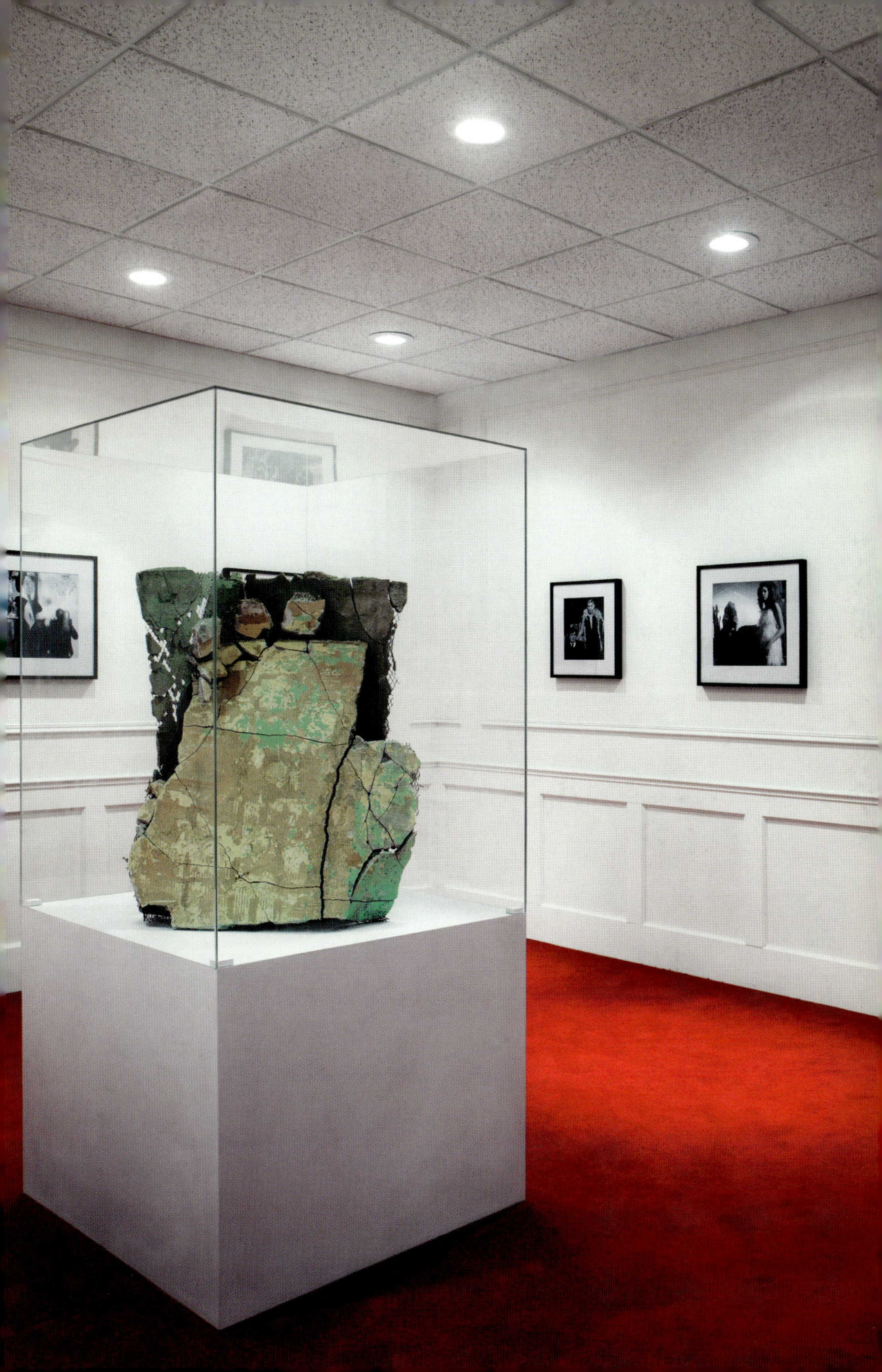

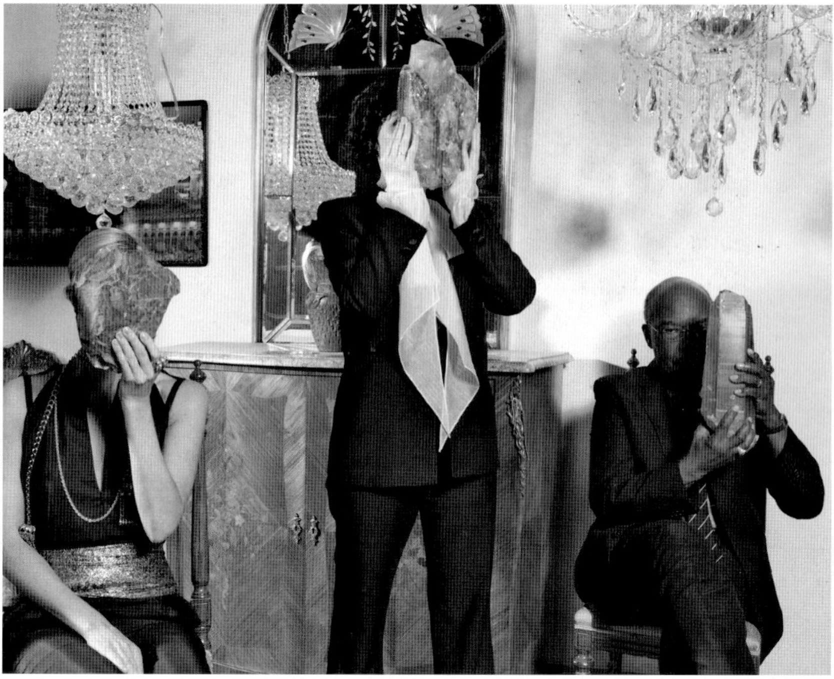

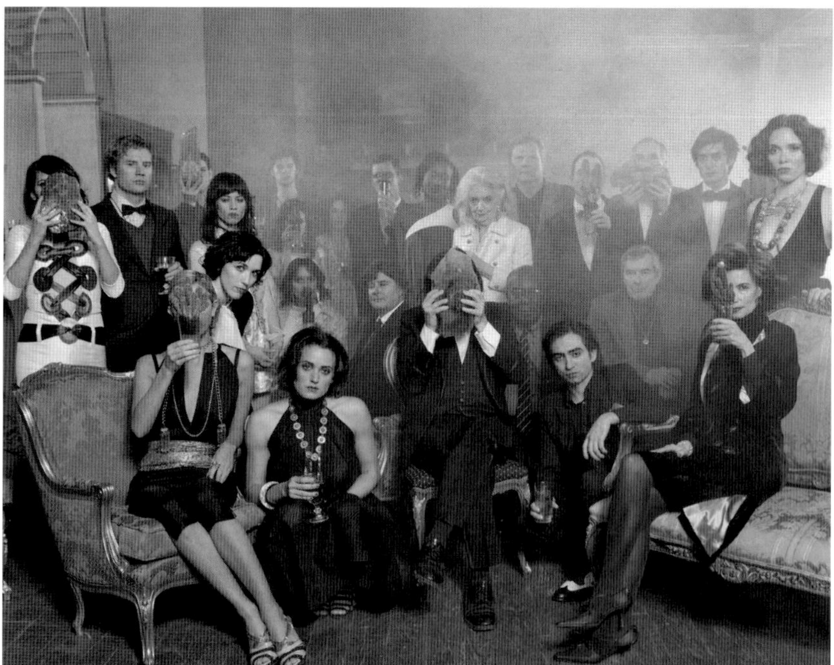

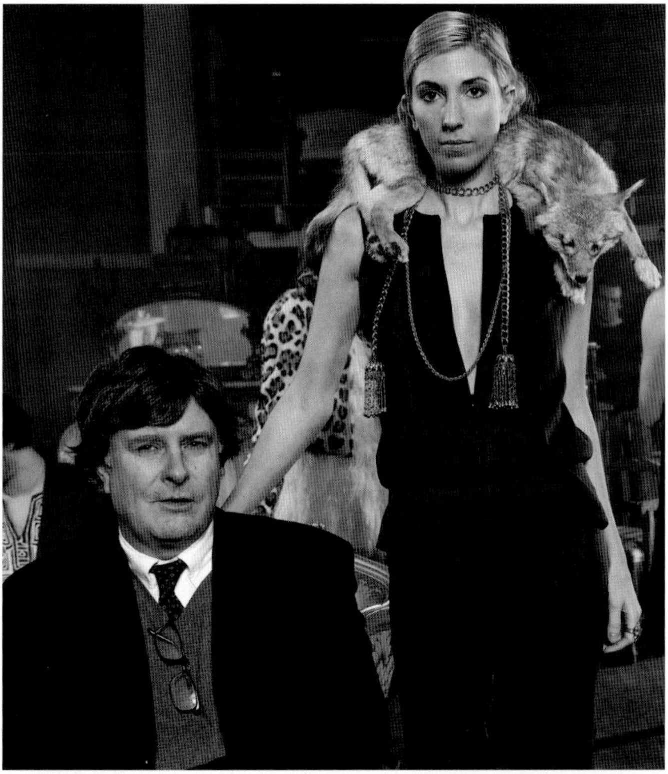

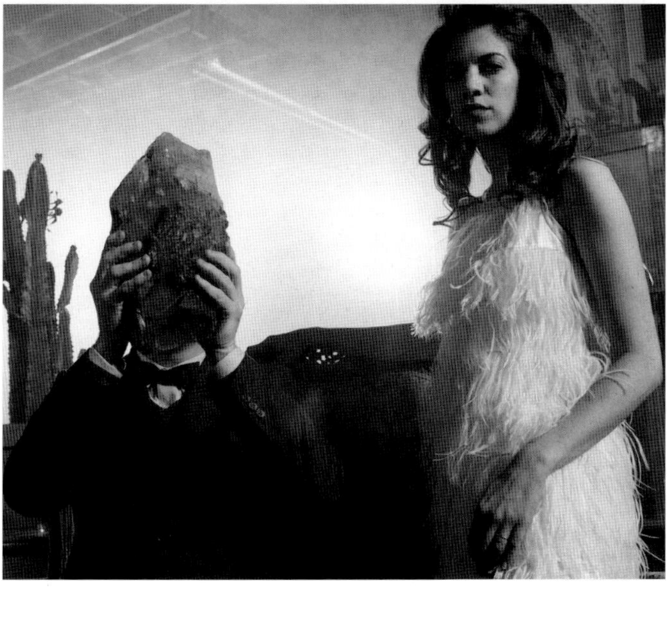

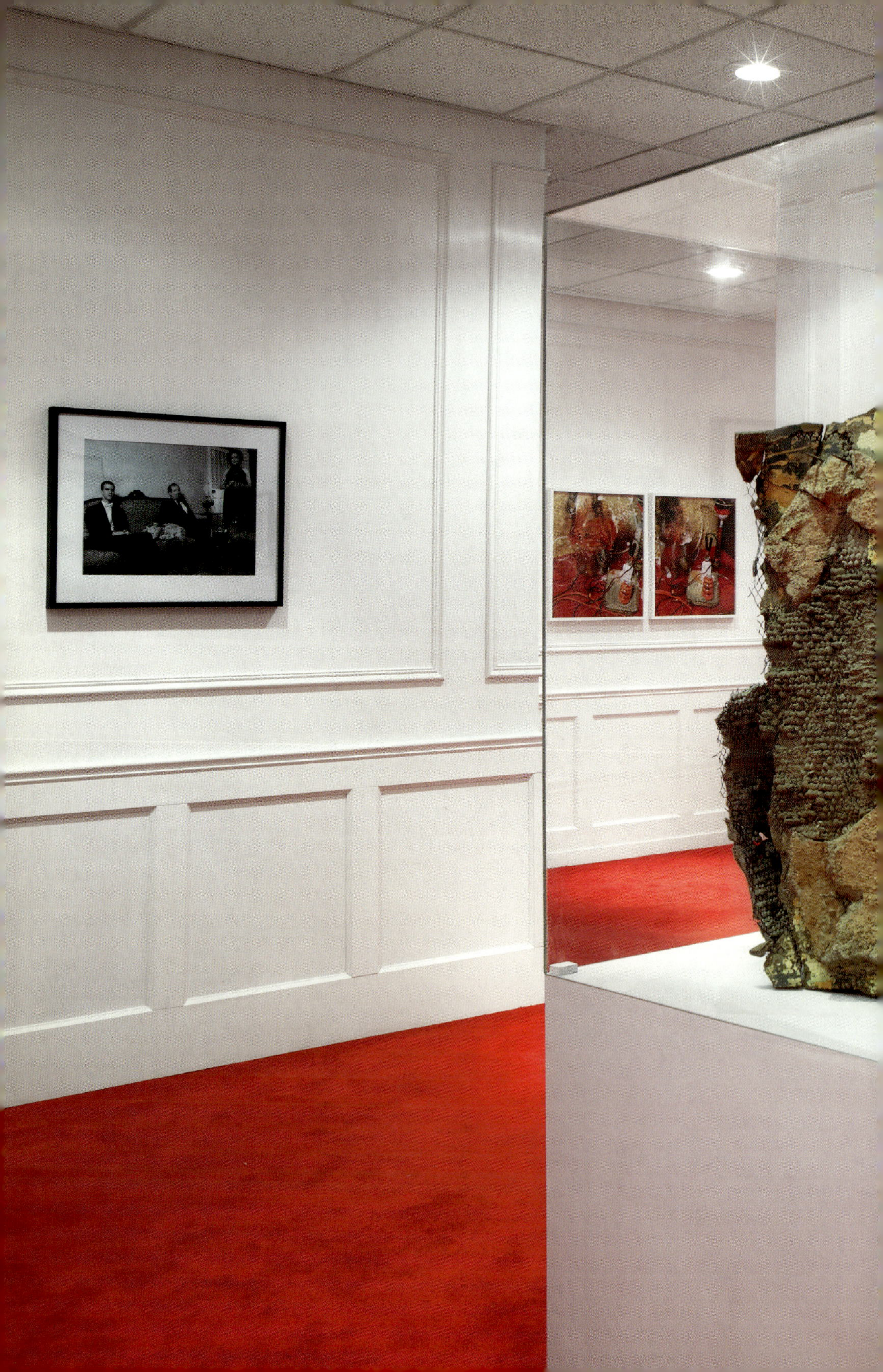

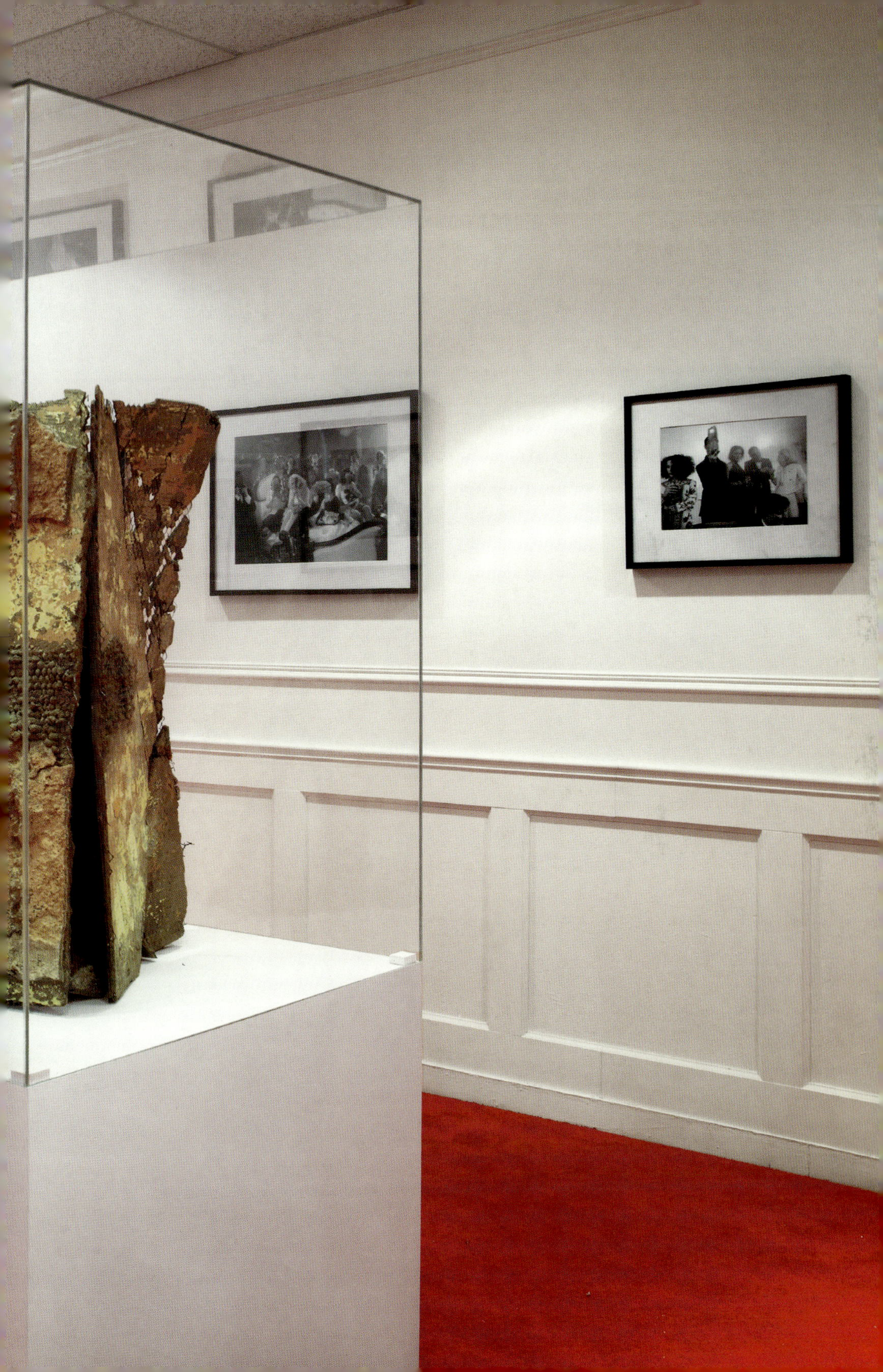

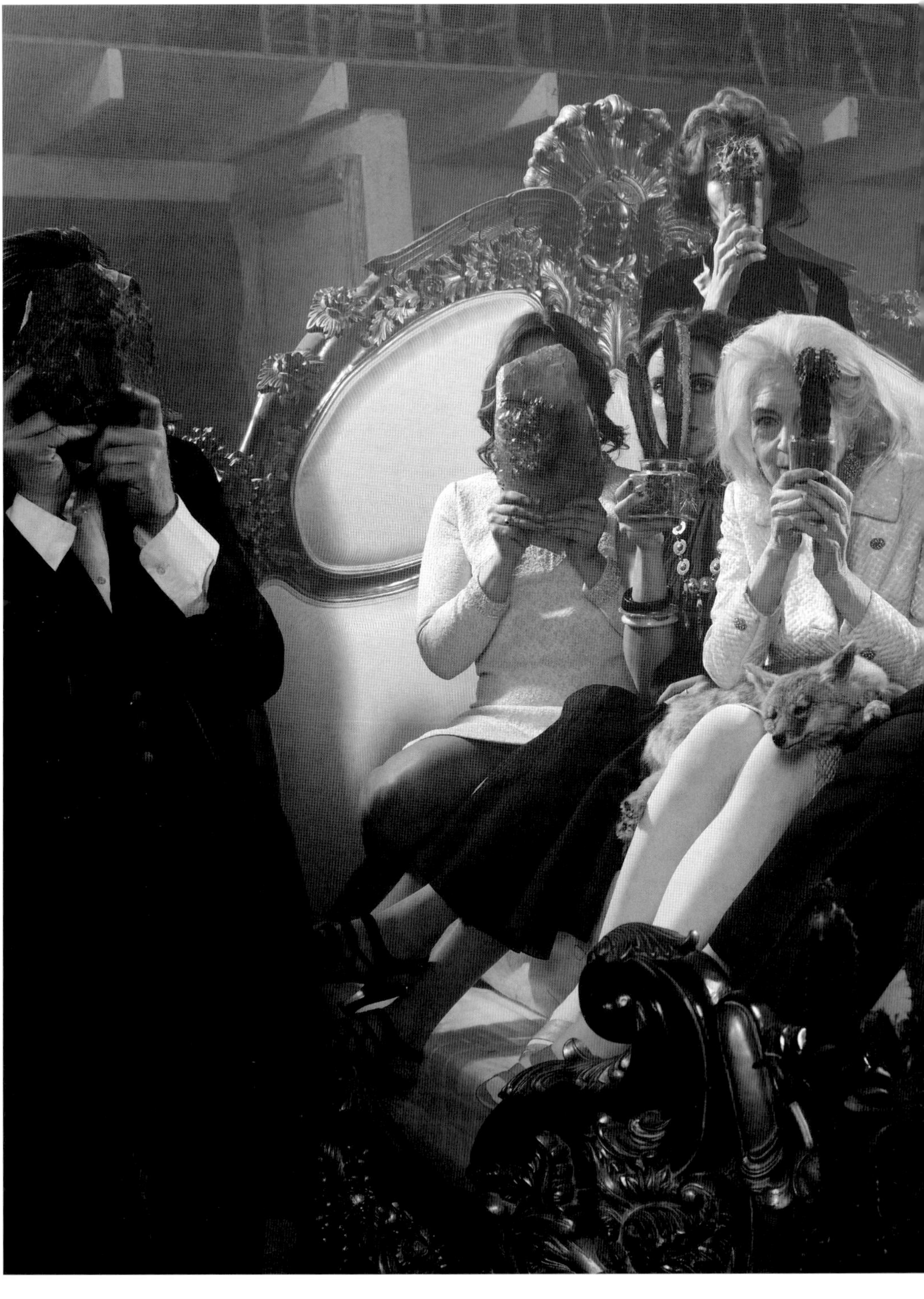

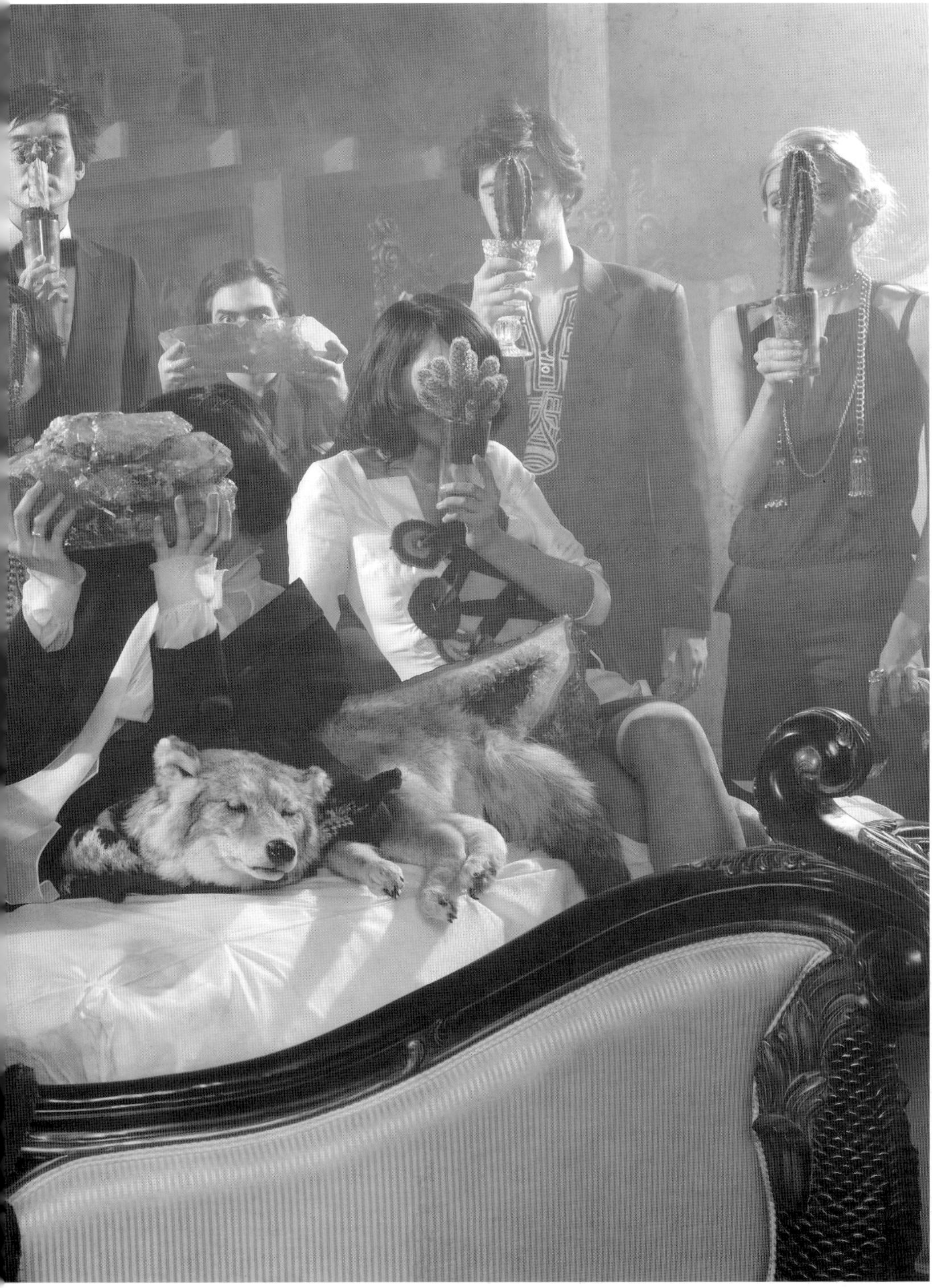

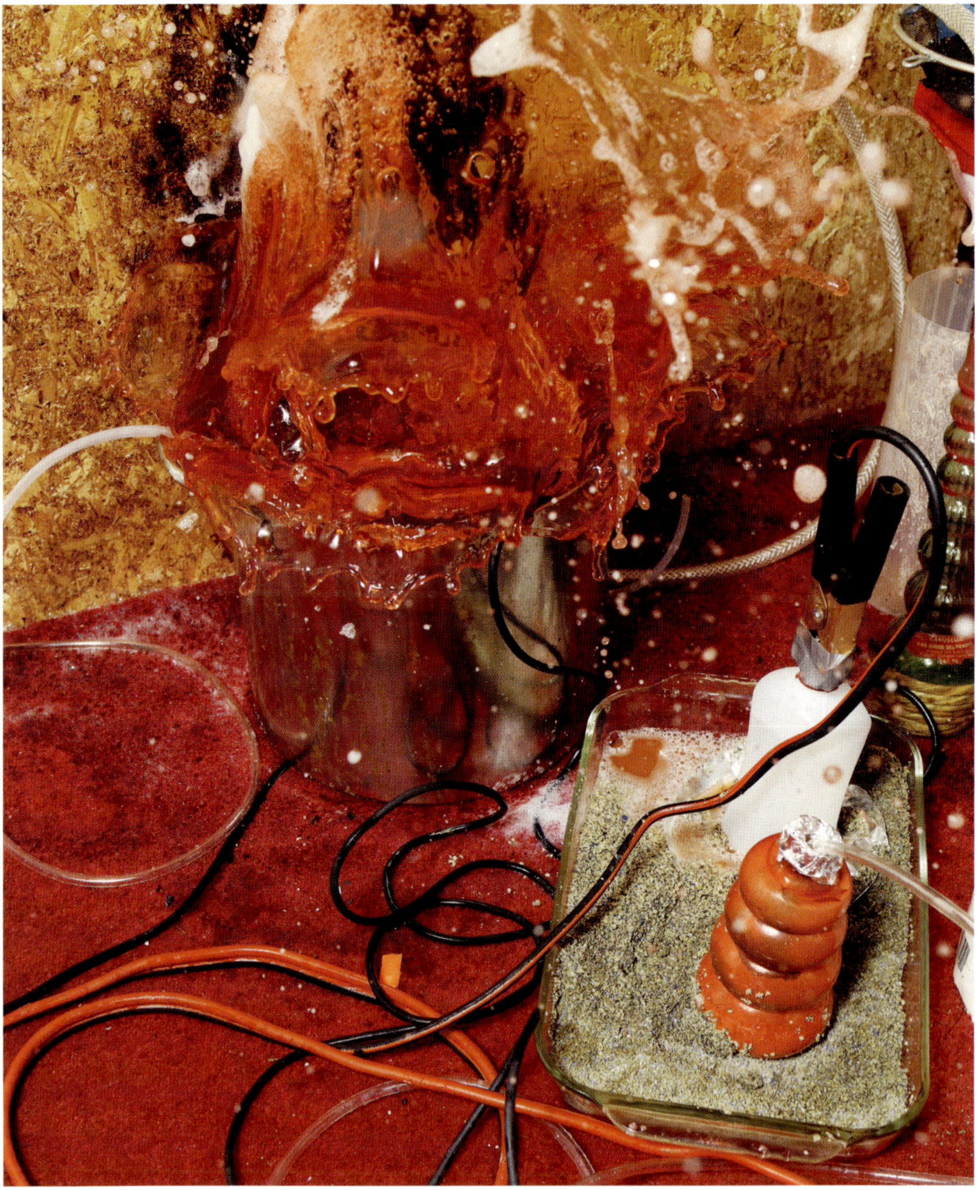

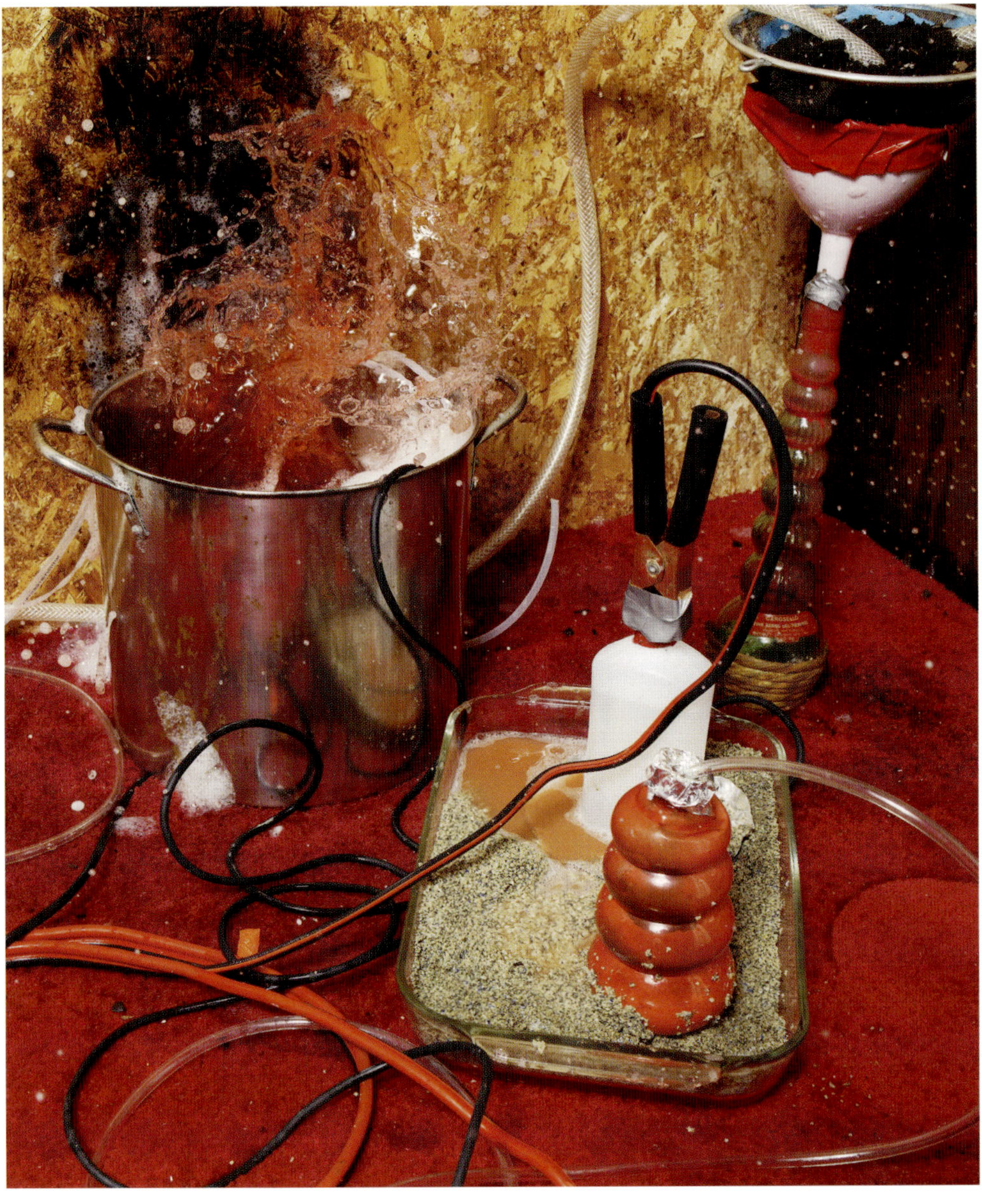

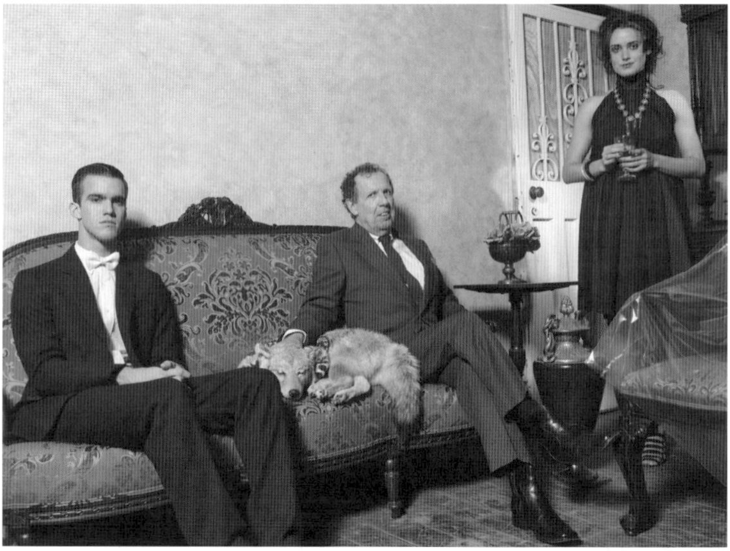

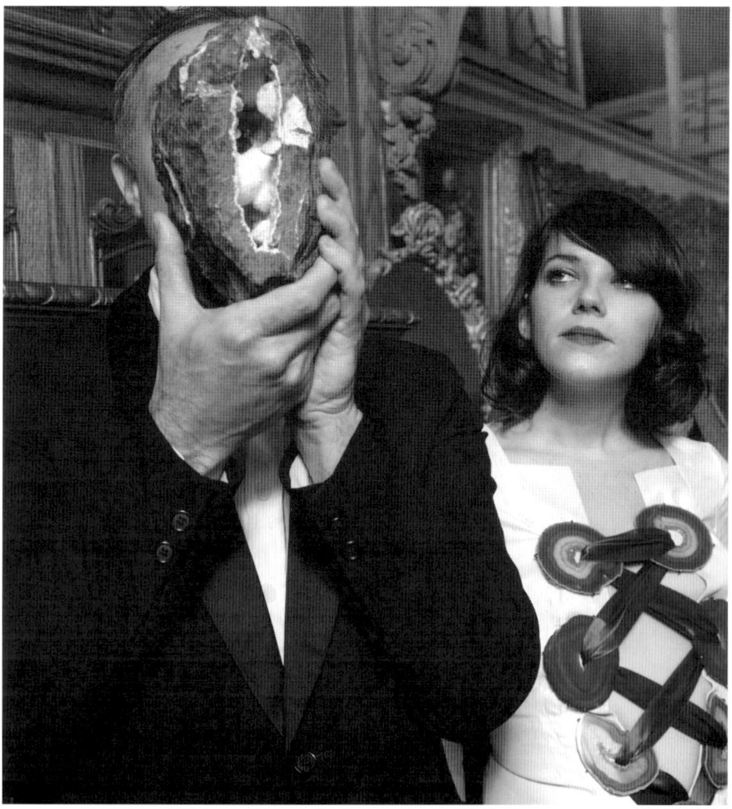

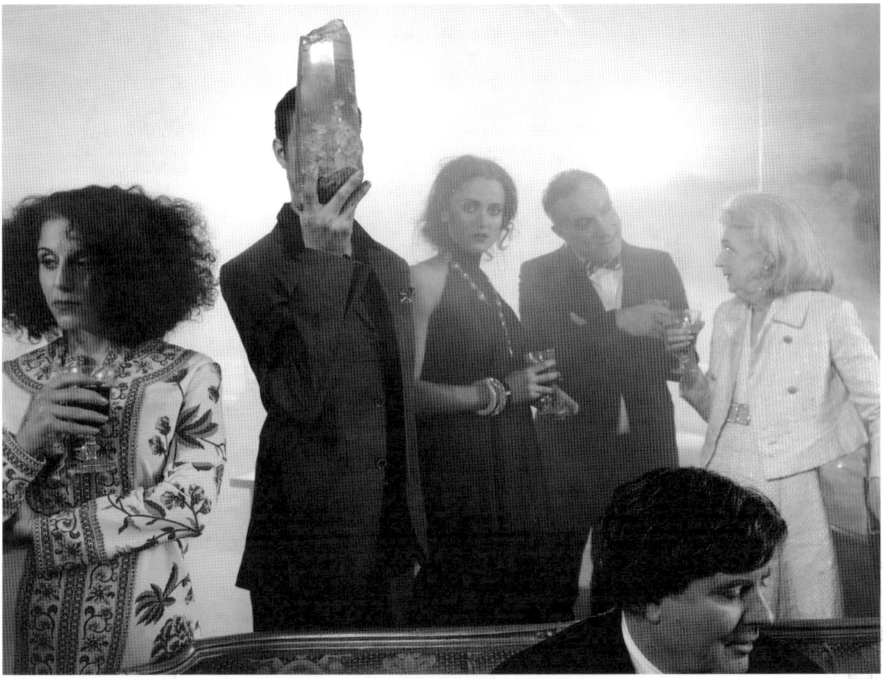

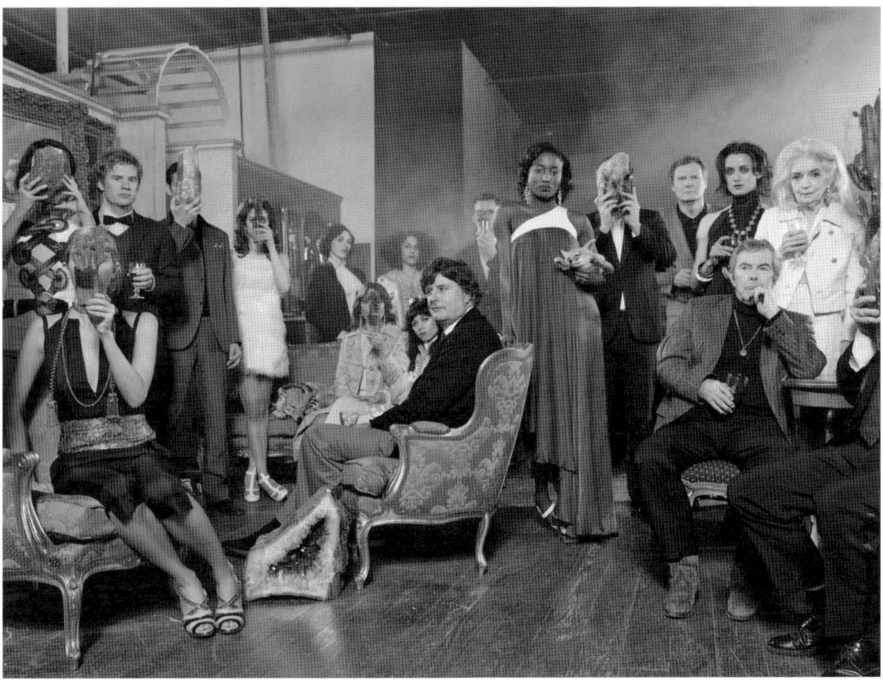

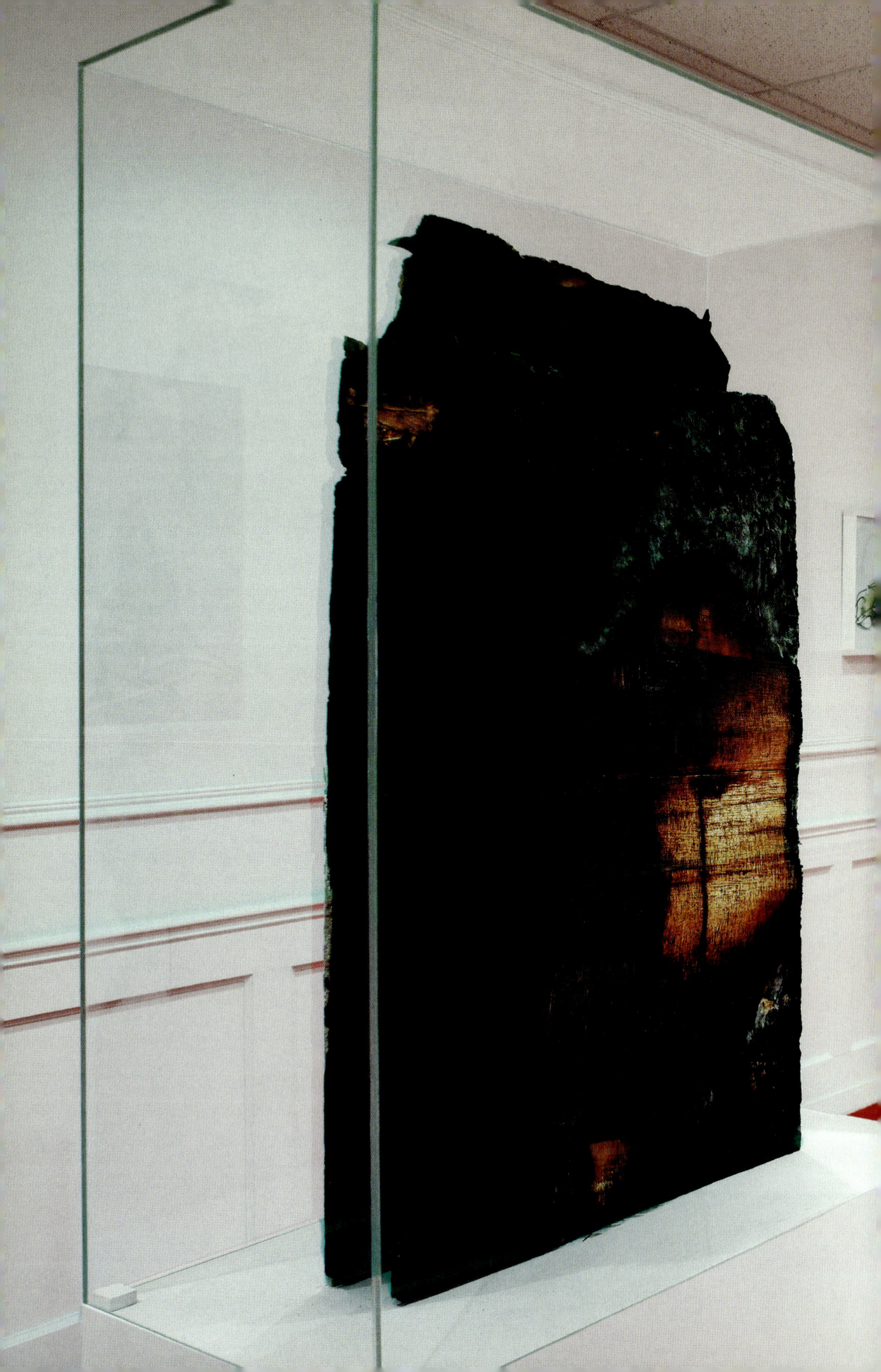

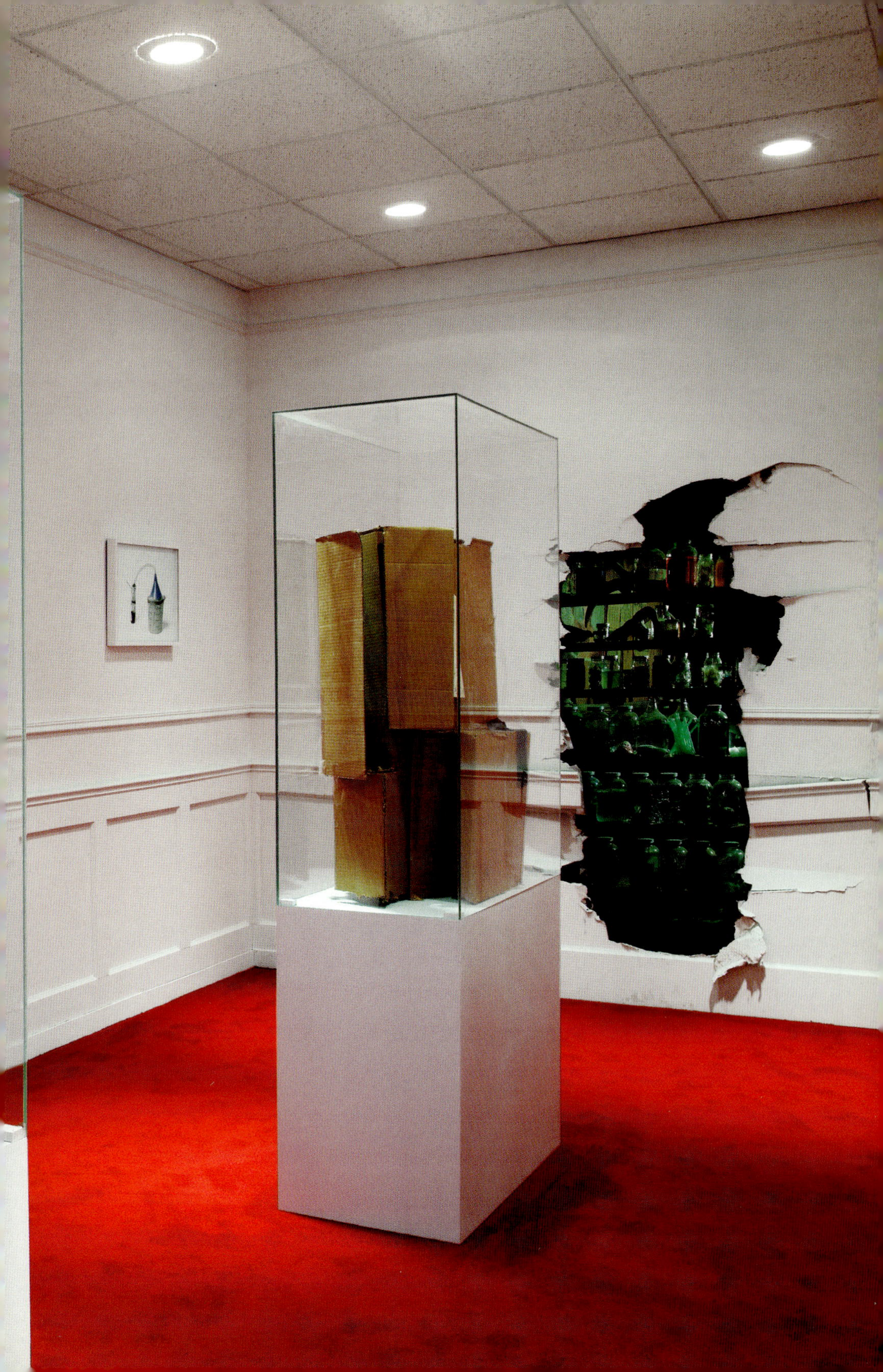

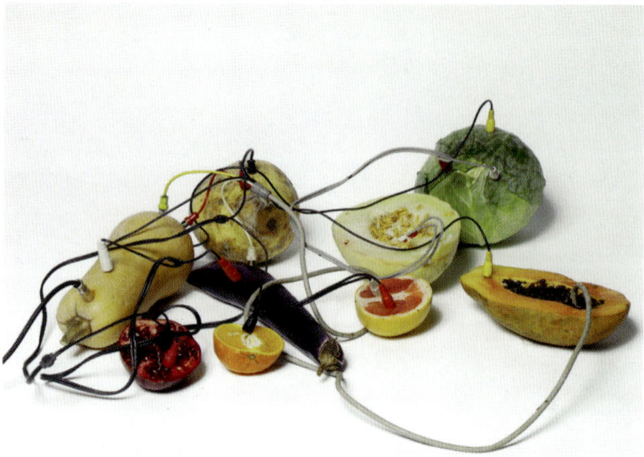

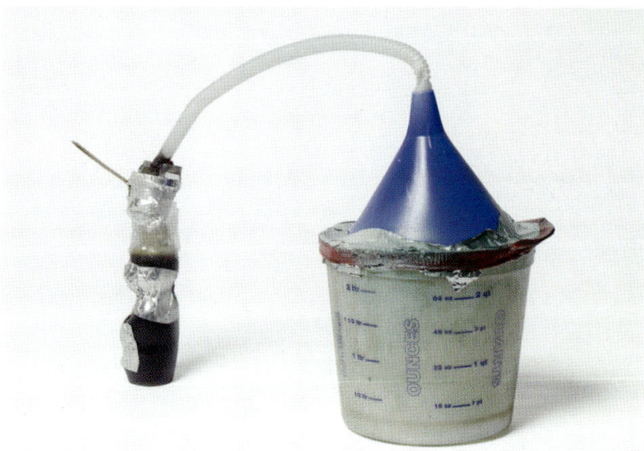
OUNCES

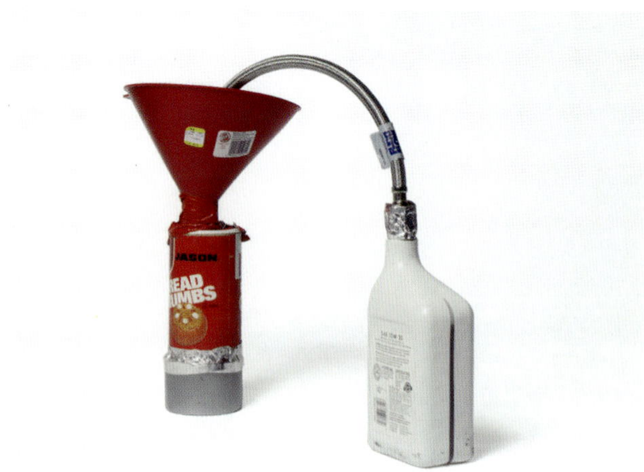
JASON

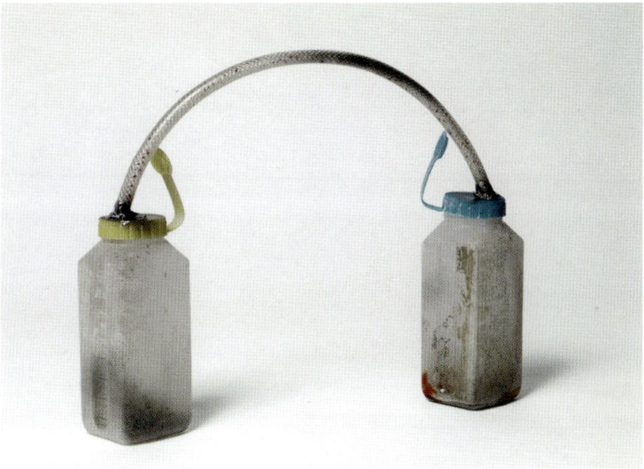

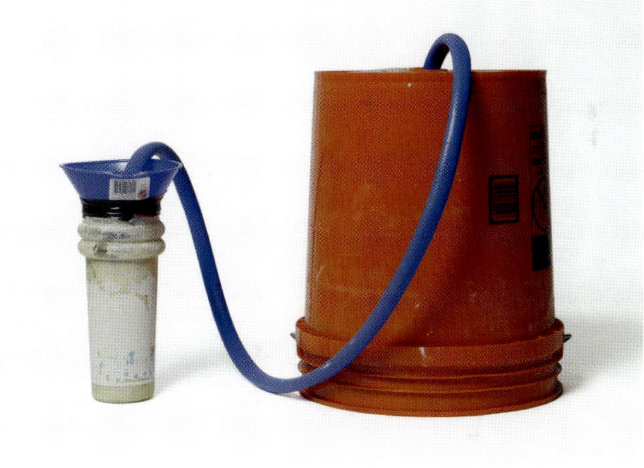

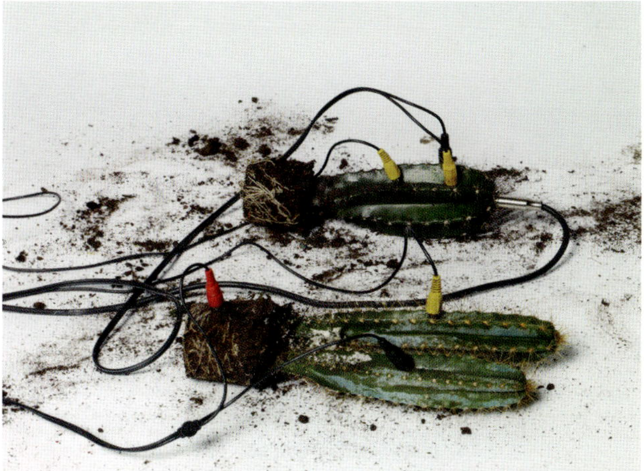

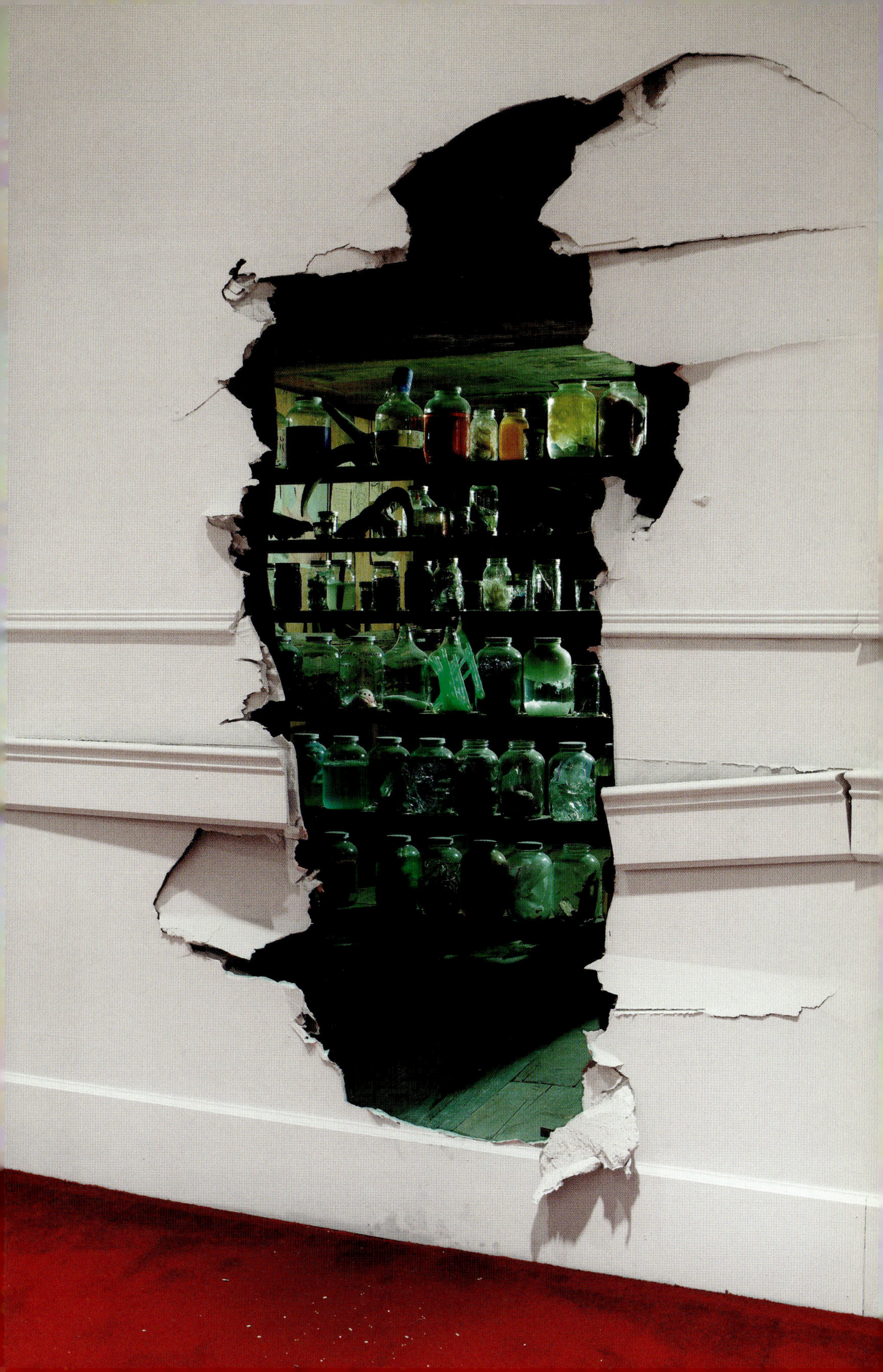

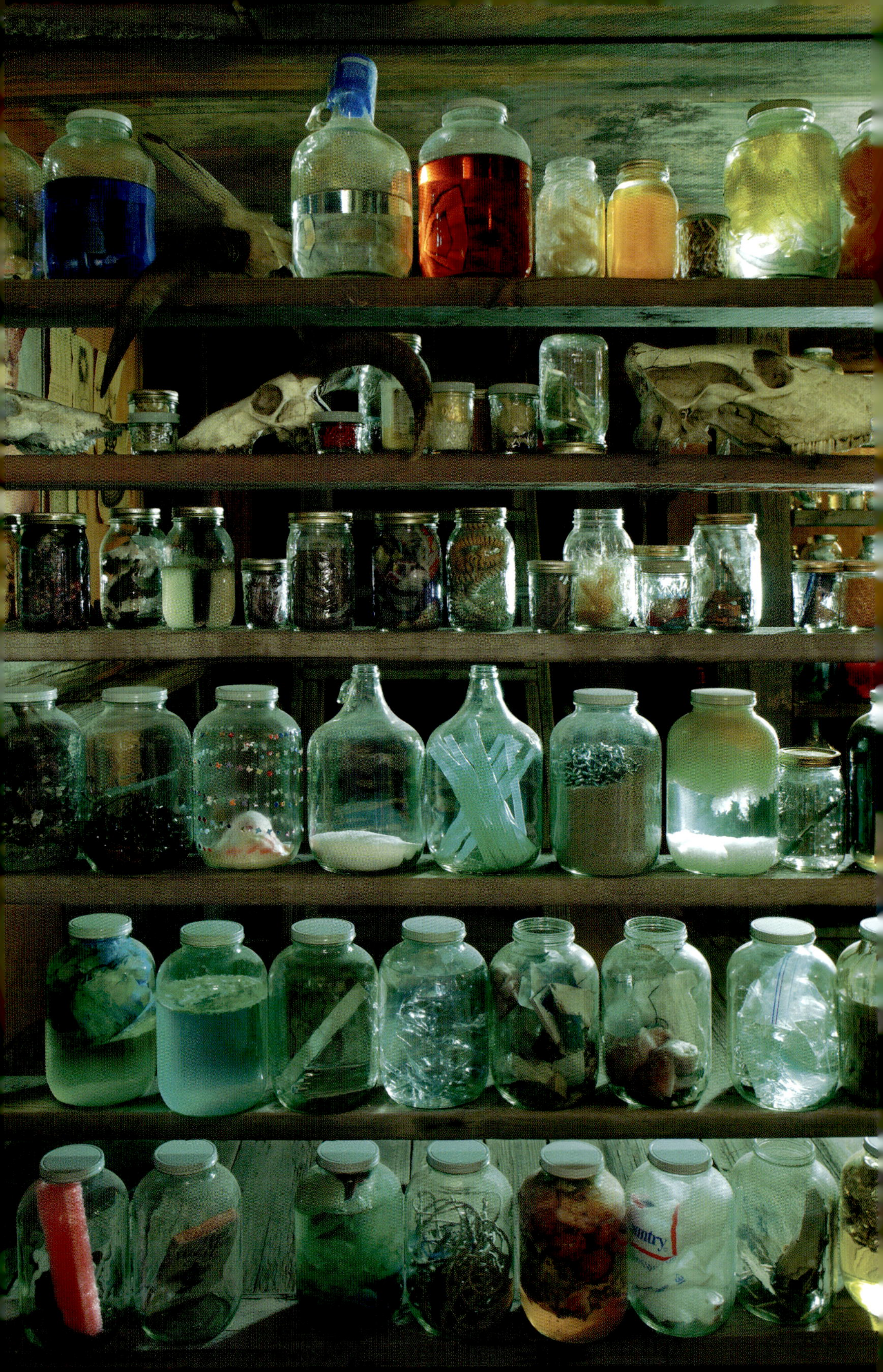

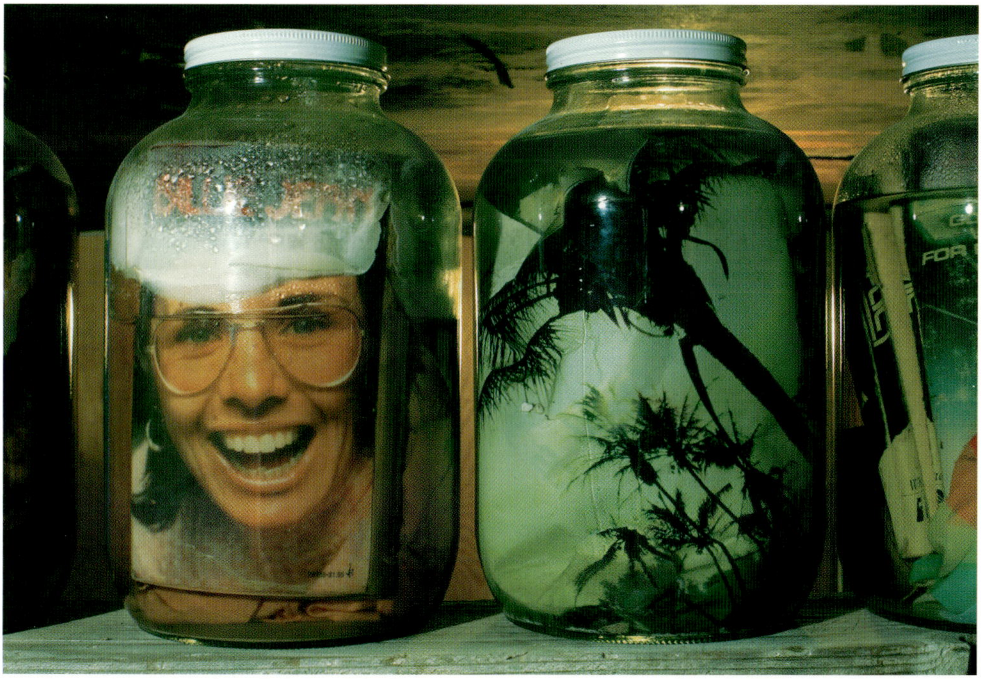

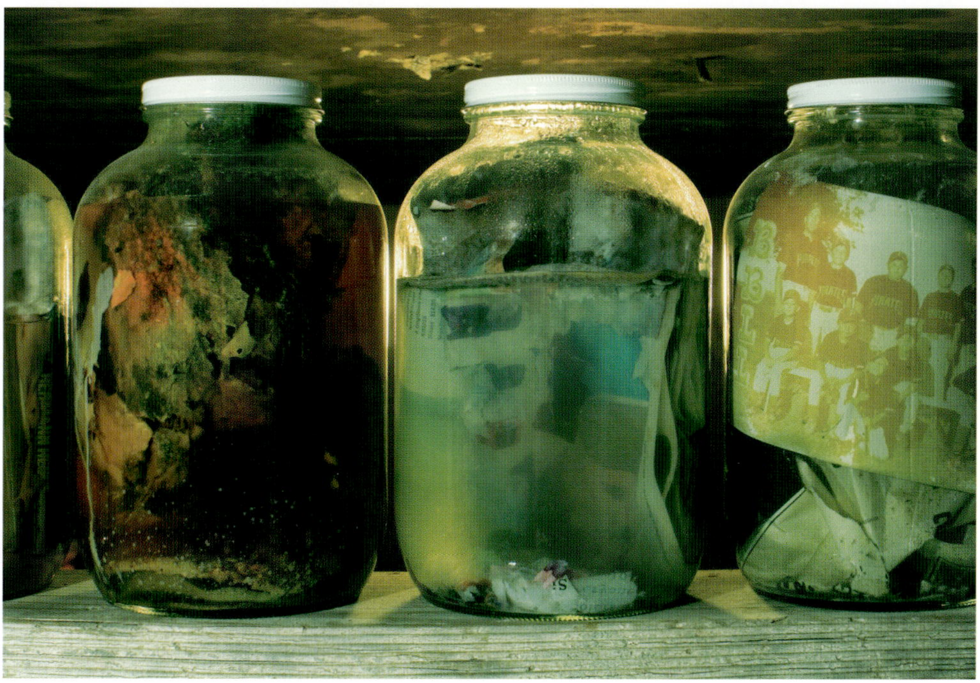

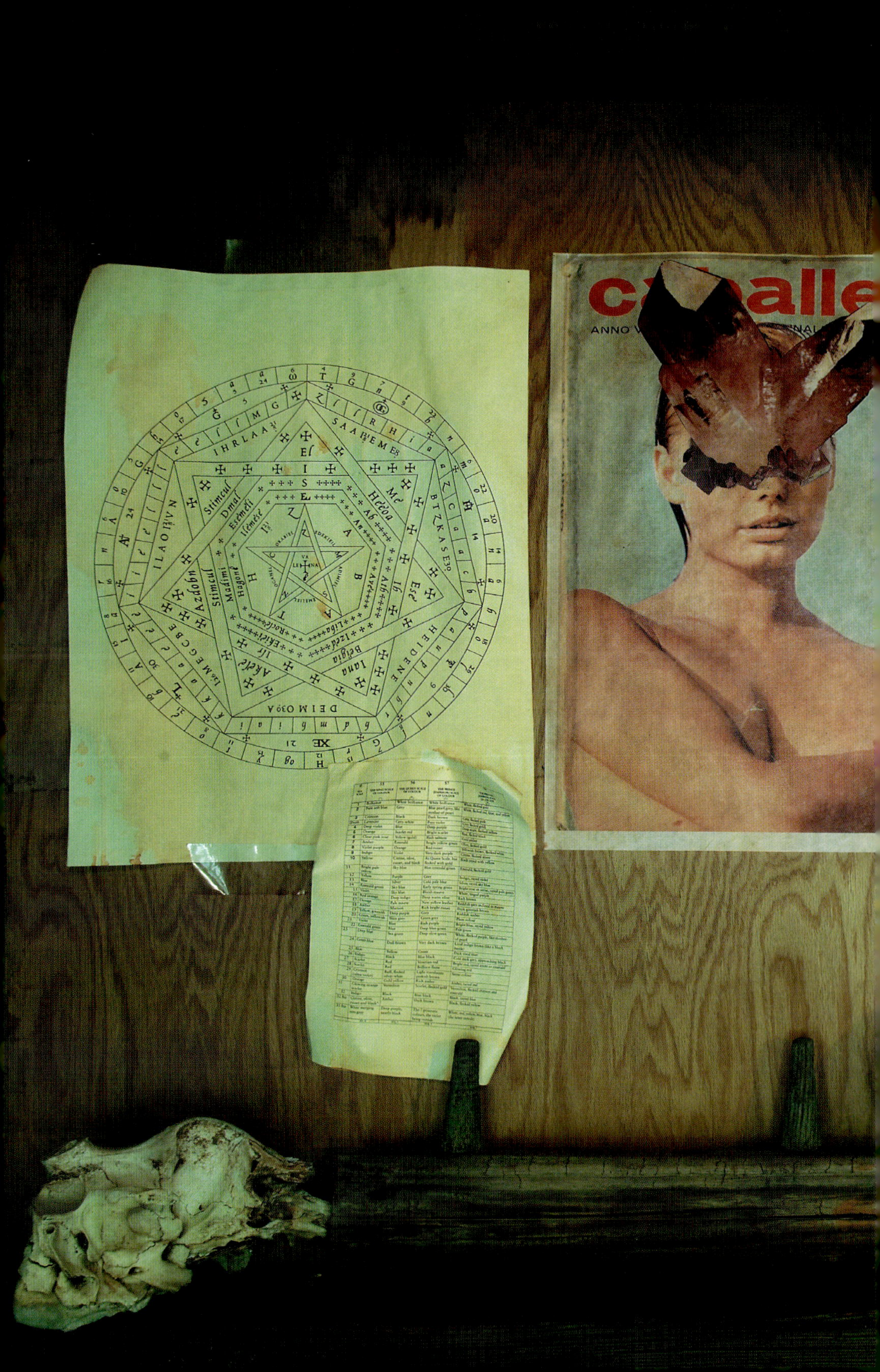

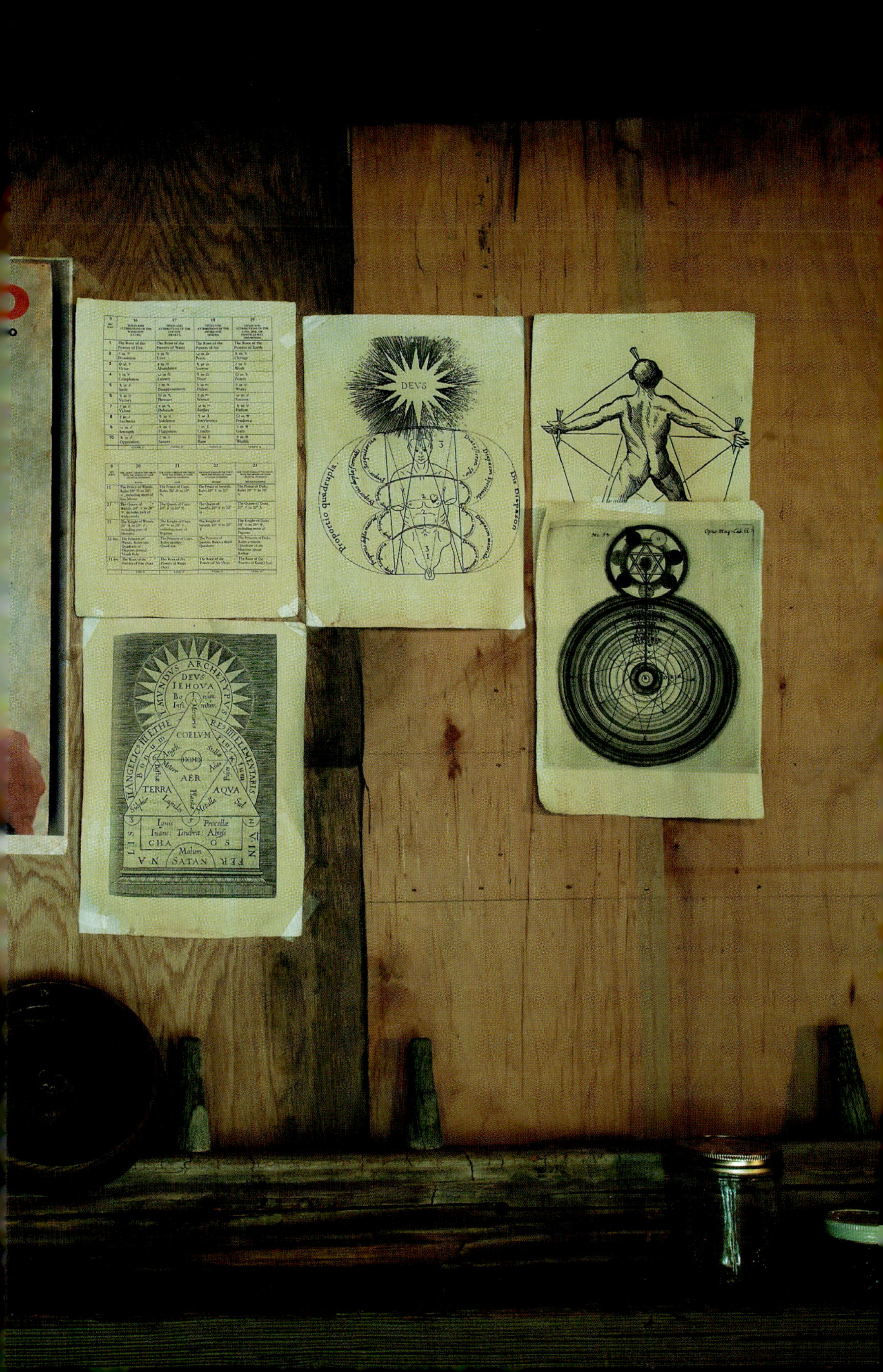
DEVS
Dis Diapason
Proportio quadrupla.
MVNDVS ARCHETYPVS
DEVS
JEHOVA
COELVM
HOMO
AER
TERRA
AQVA
Ignis
Procellæ
Inane
Tenebræ
Abyss
CHA
OS
Malum
SATAN

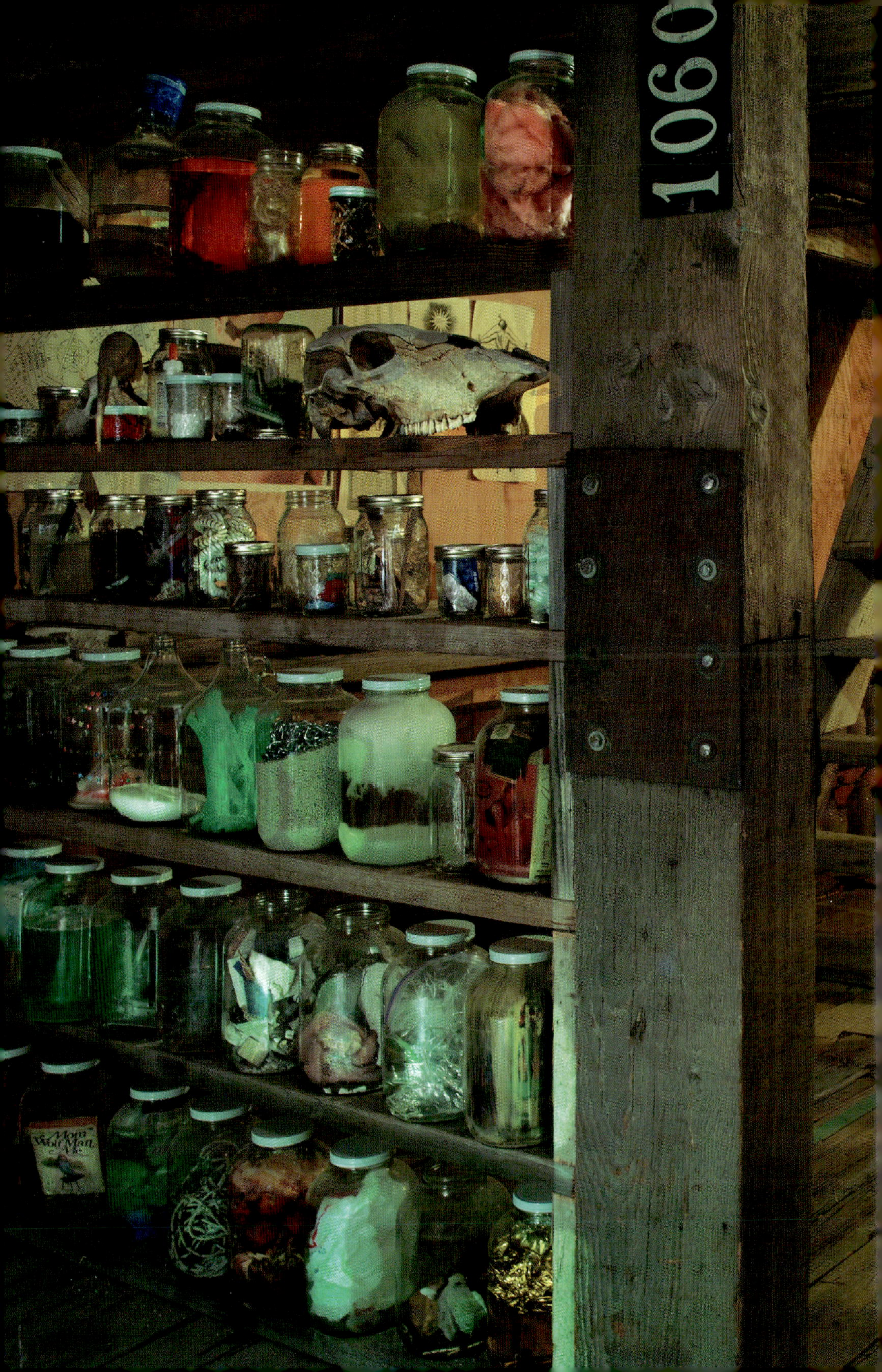
1060
Mom Wolf Man Me

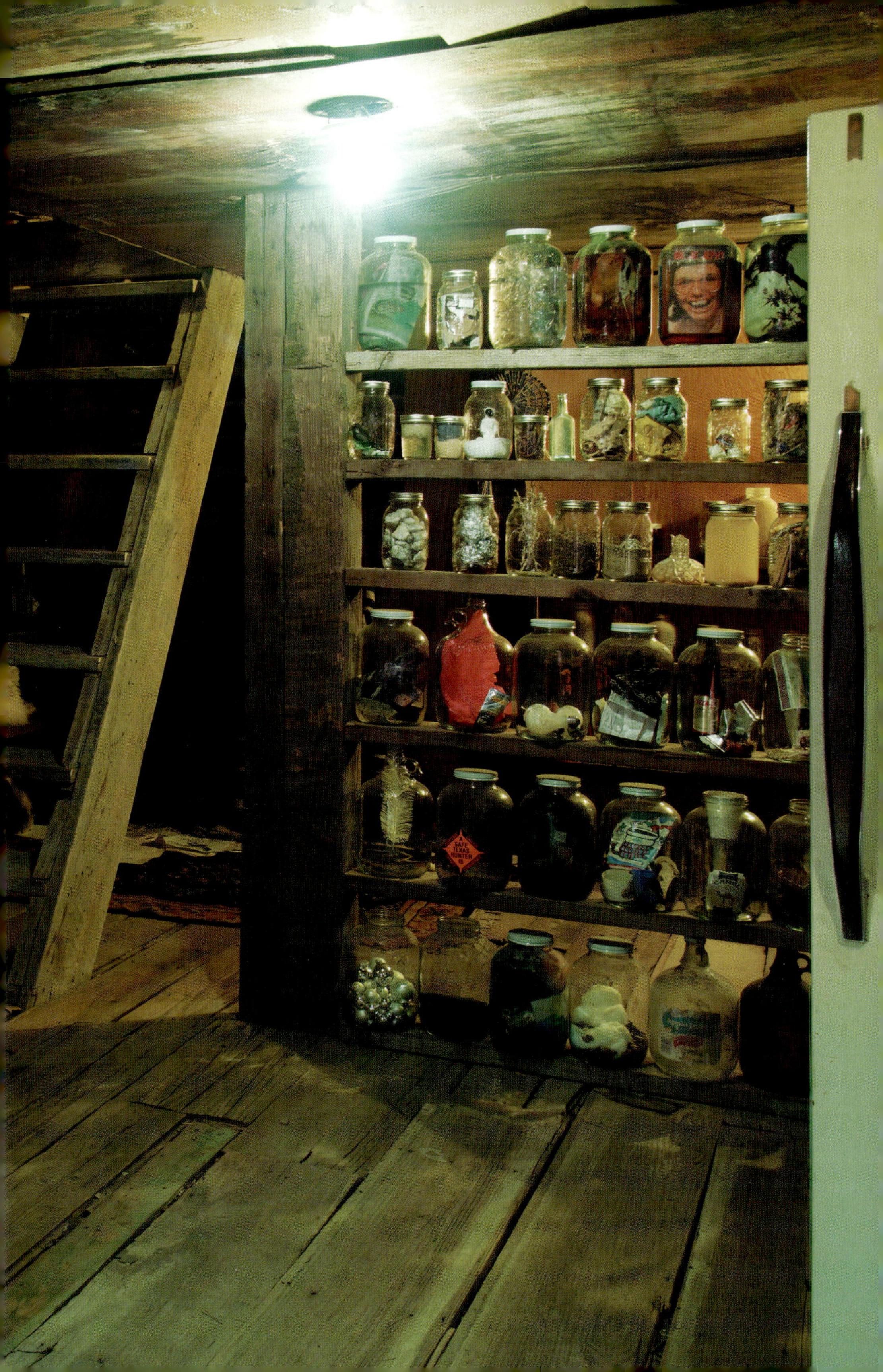
A SAFE TEXAS HUNTER IS

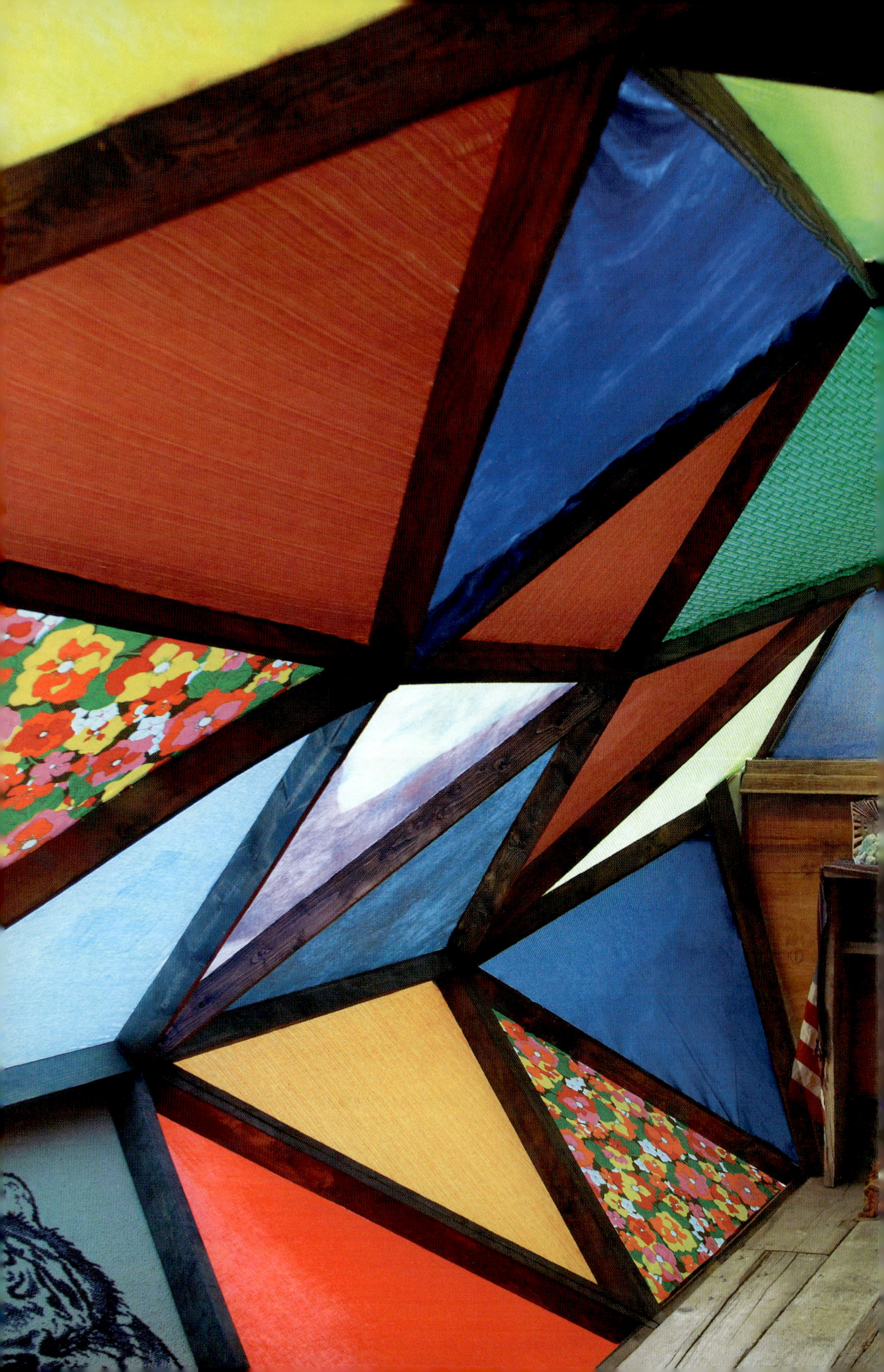

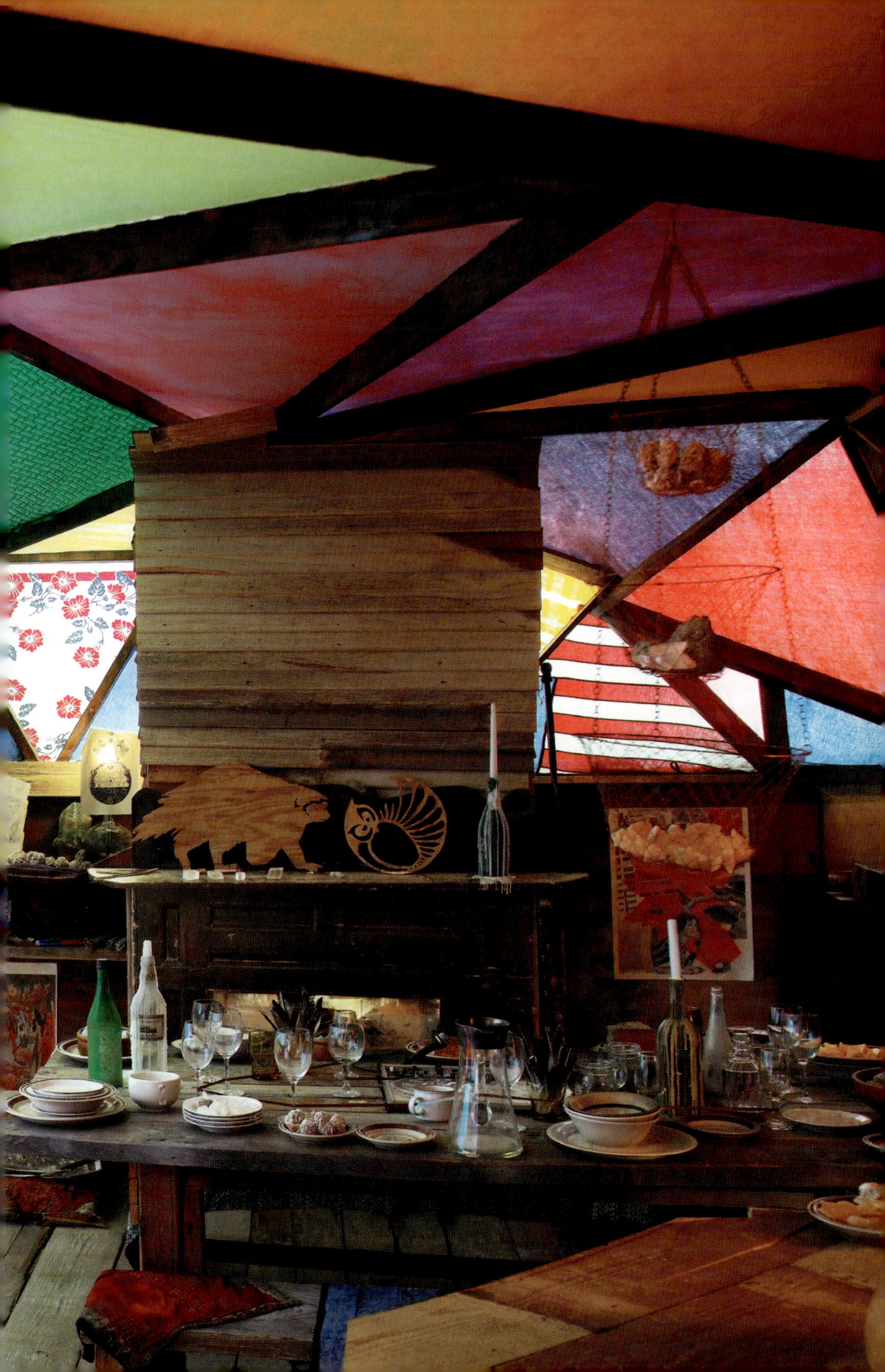

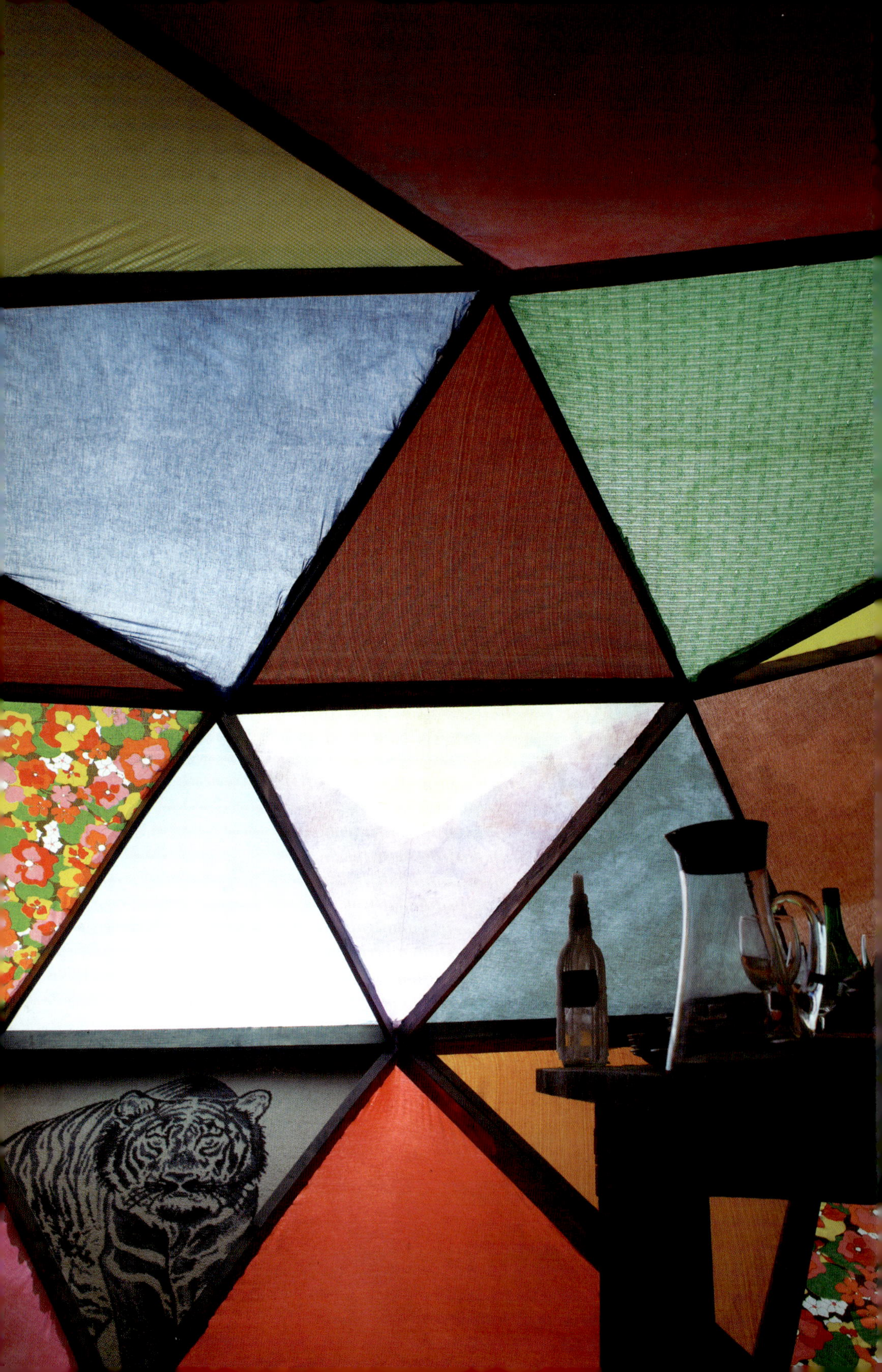

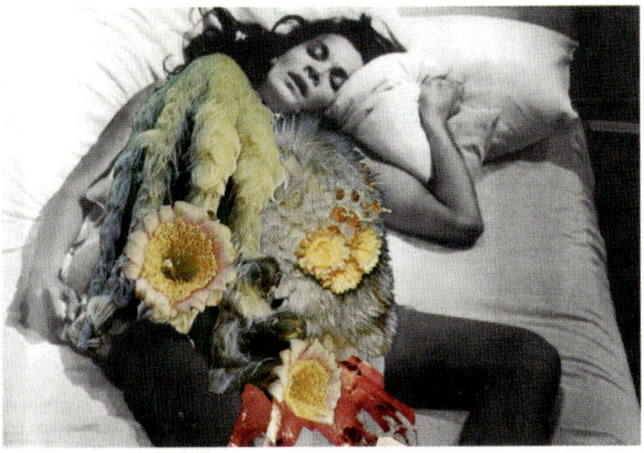

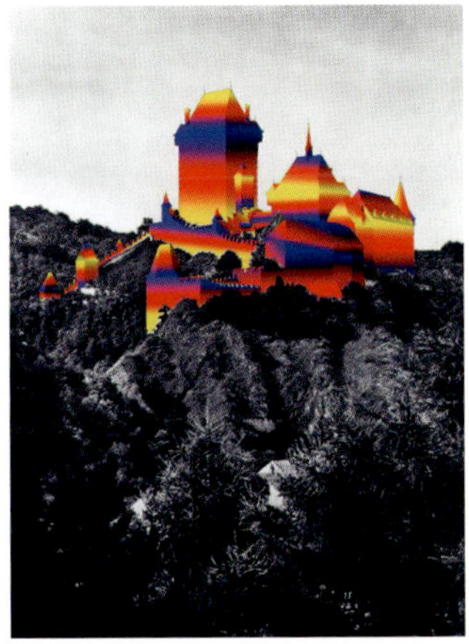

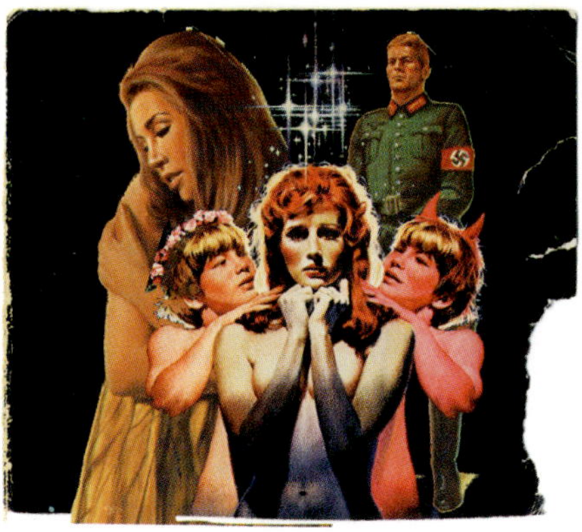

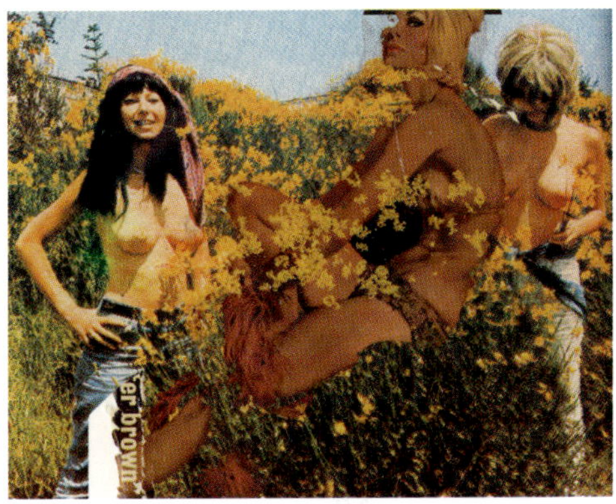
er brown

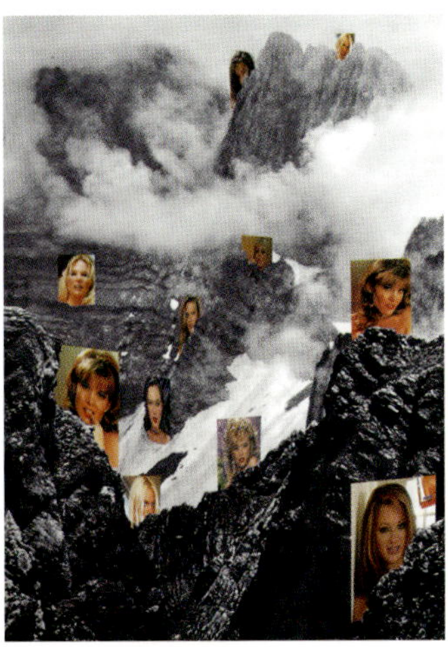

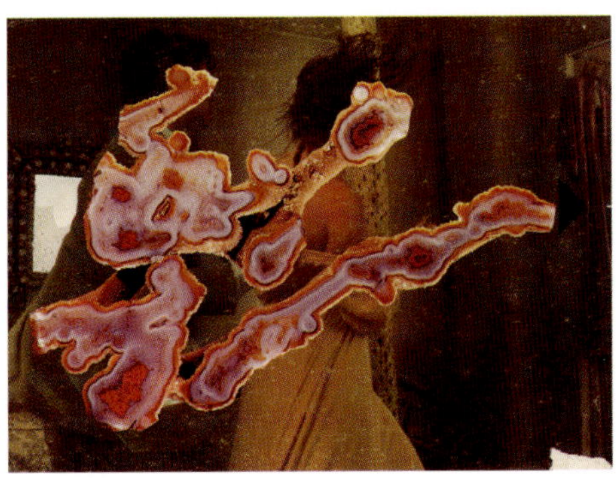

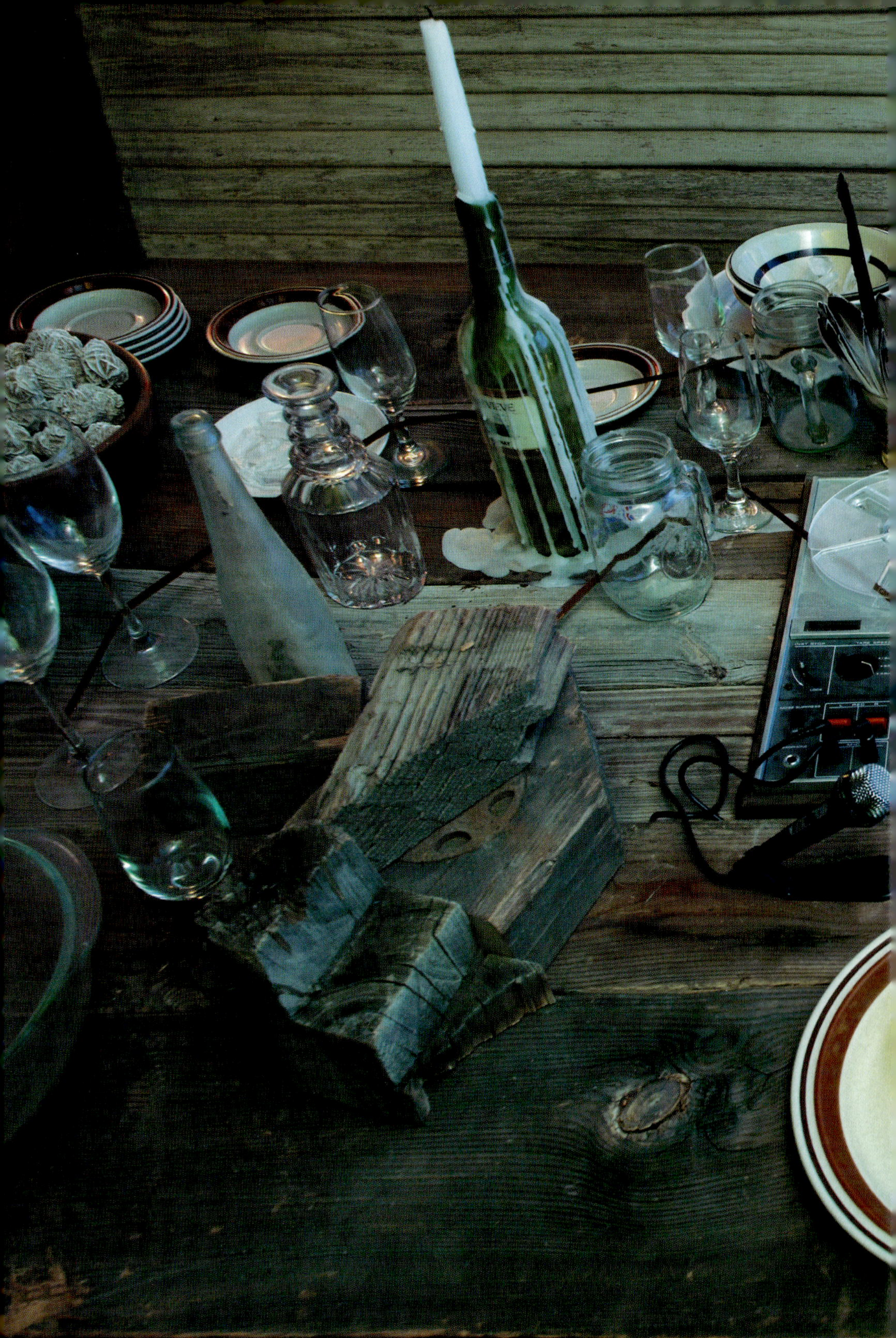

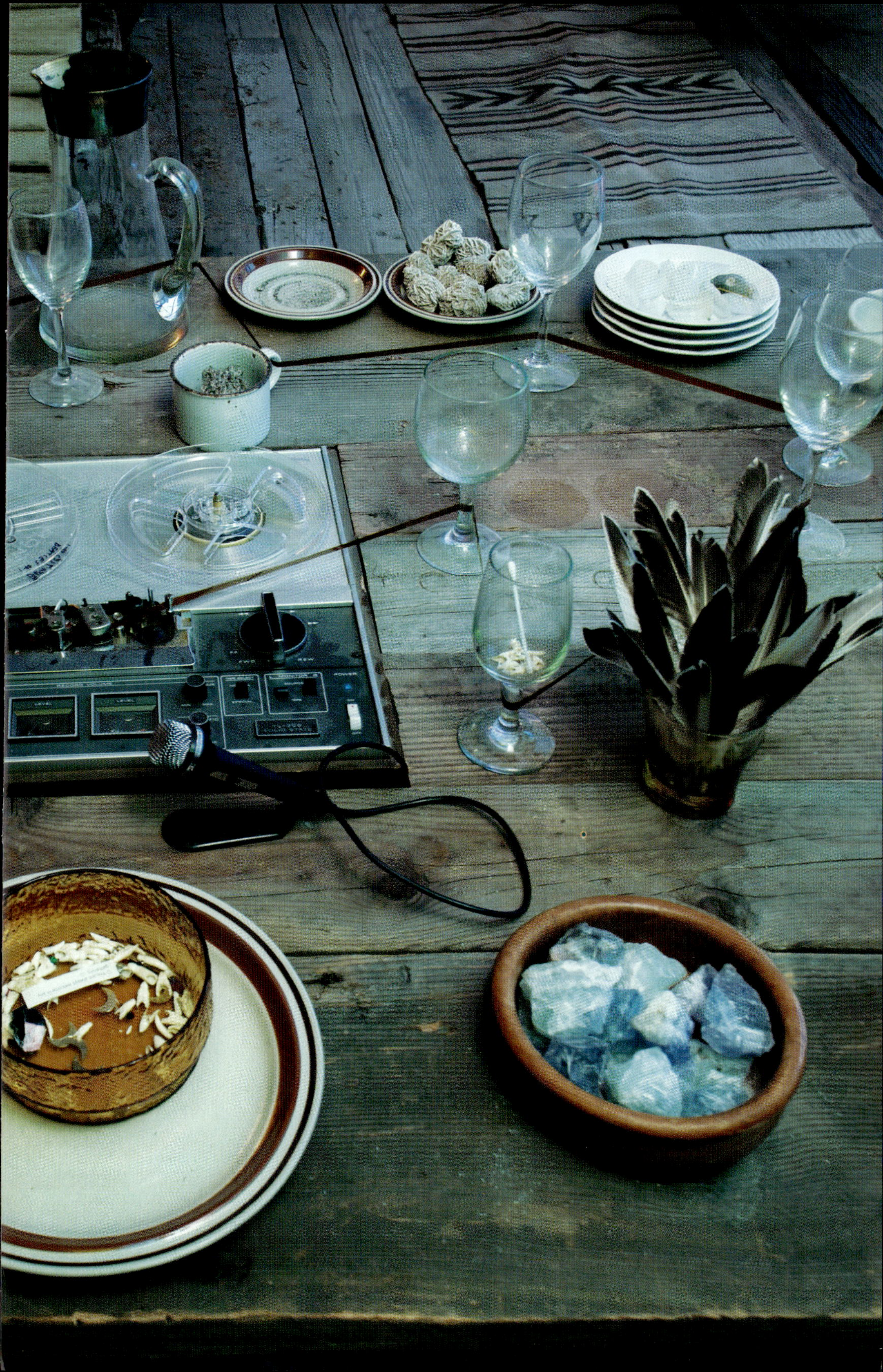

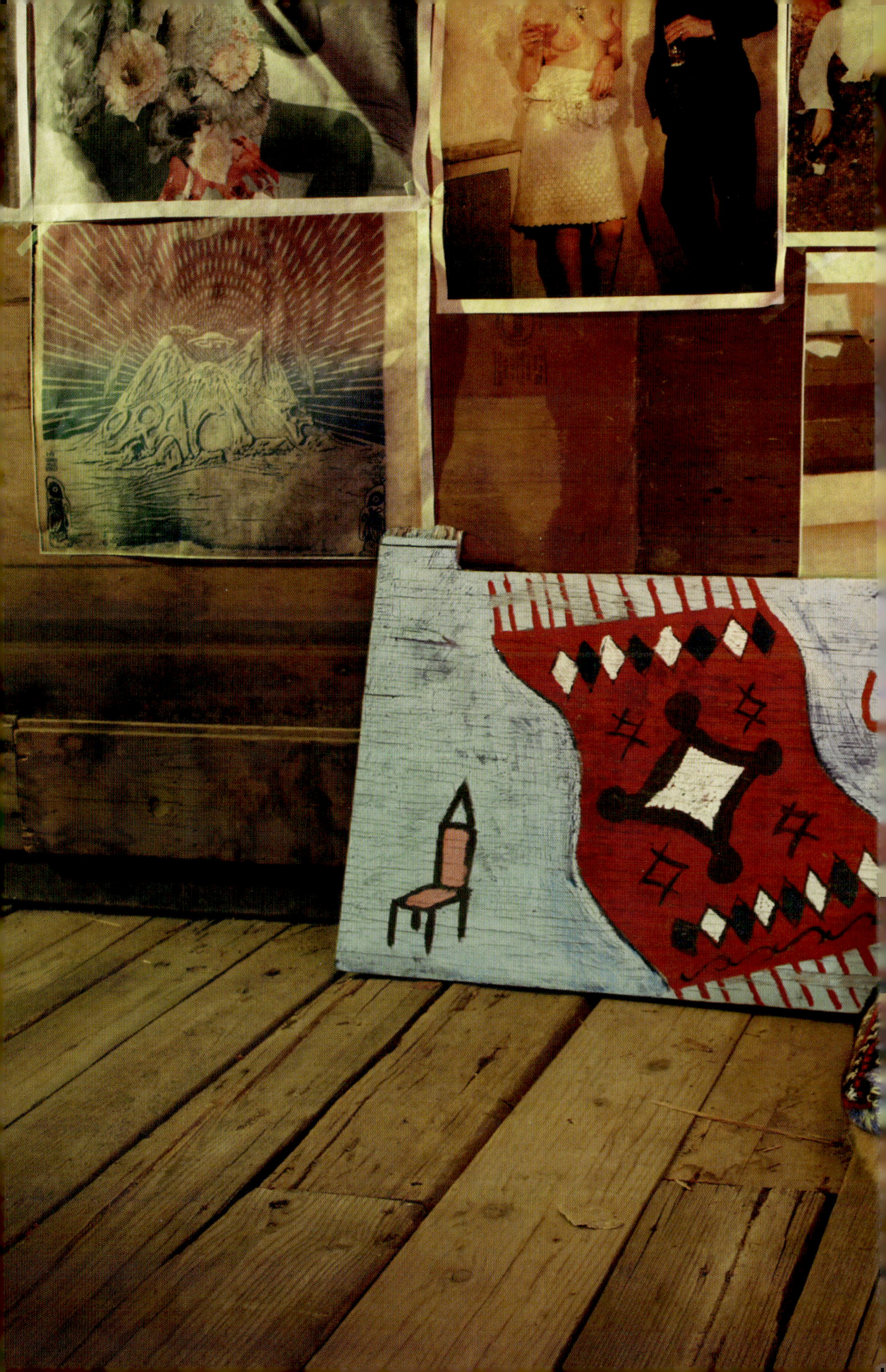

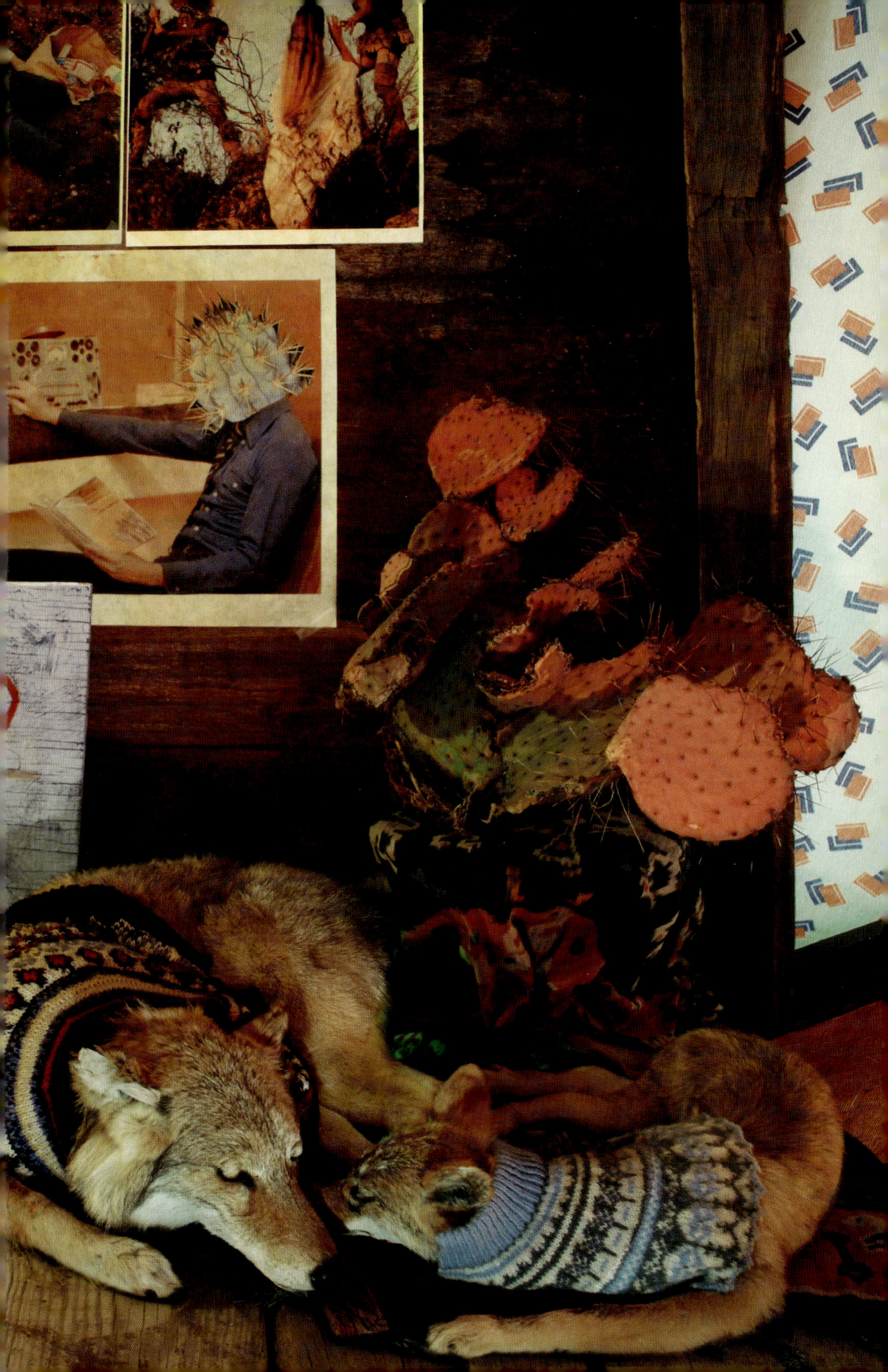

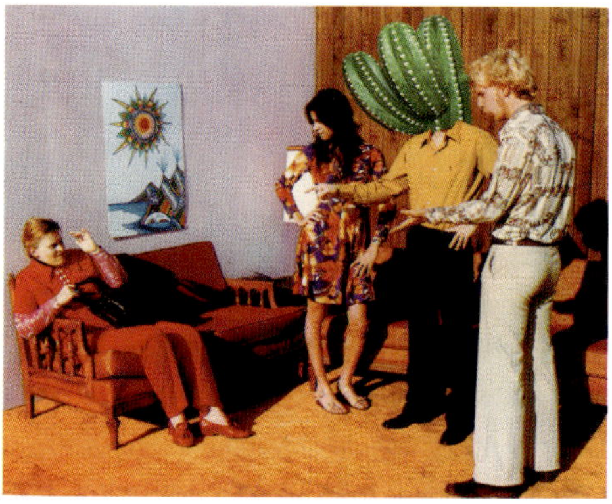

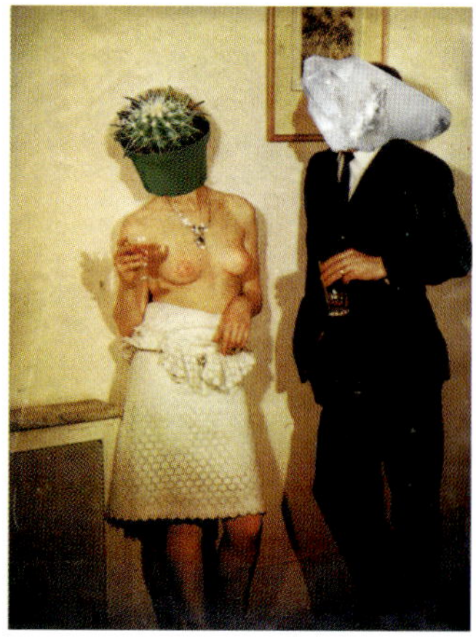

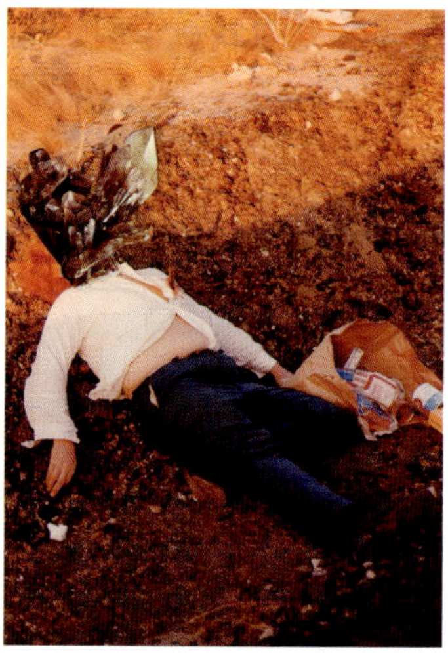

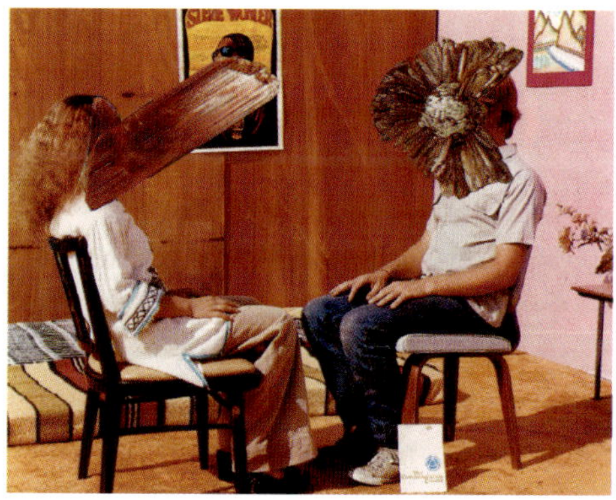

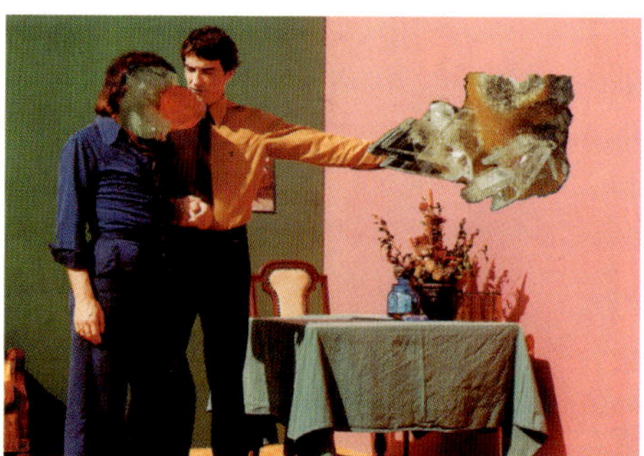

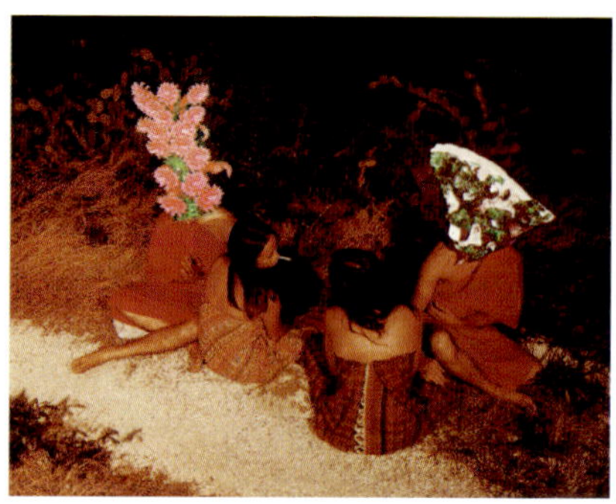

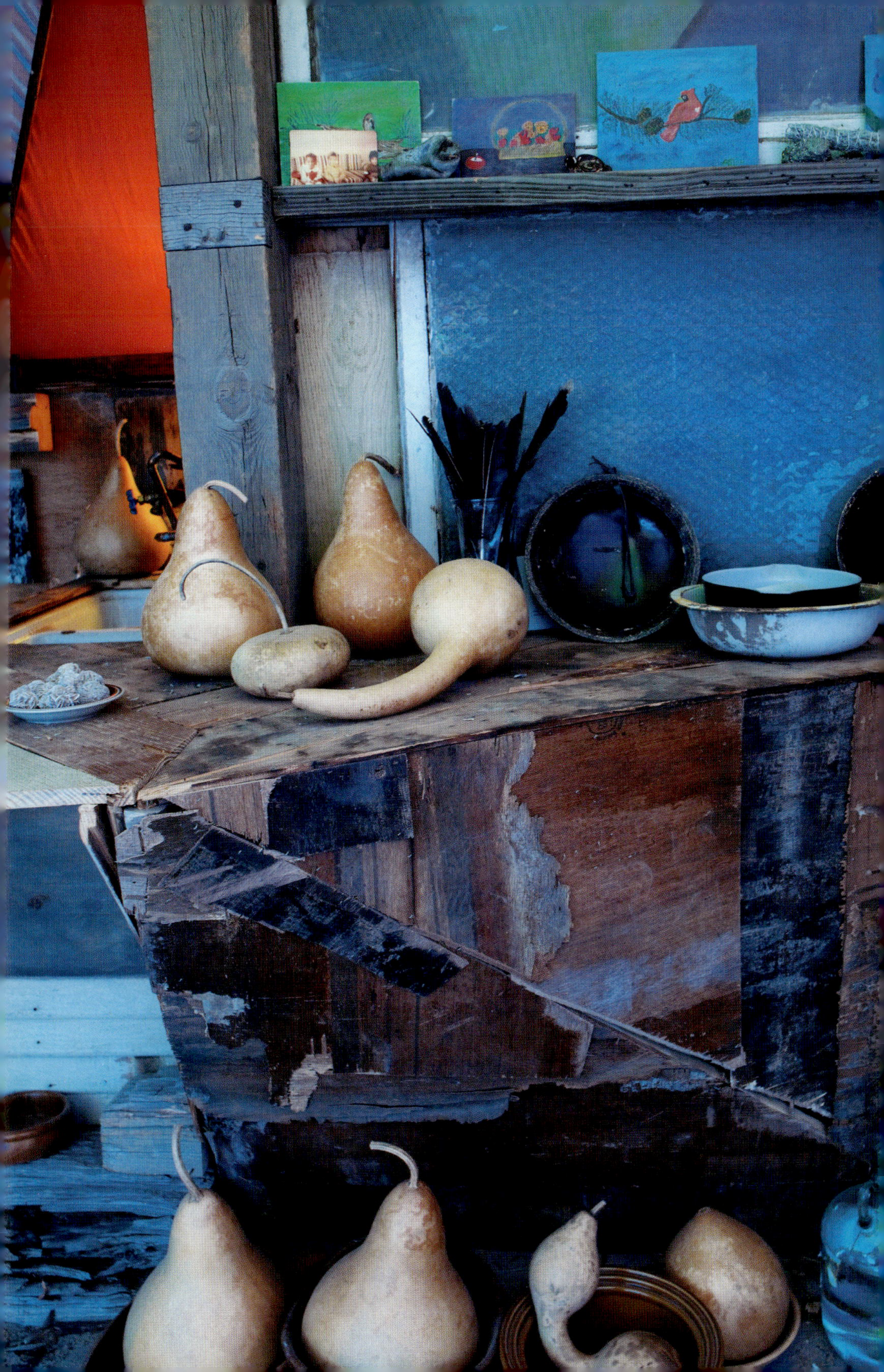

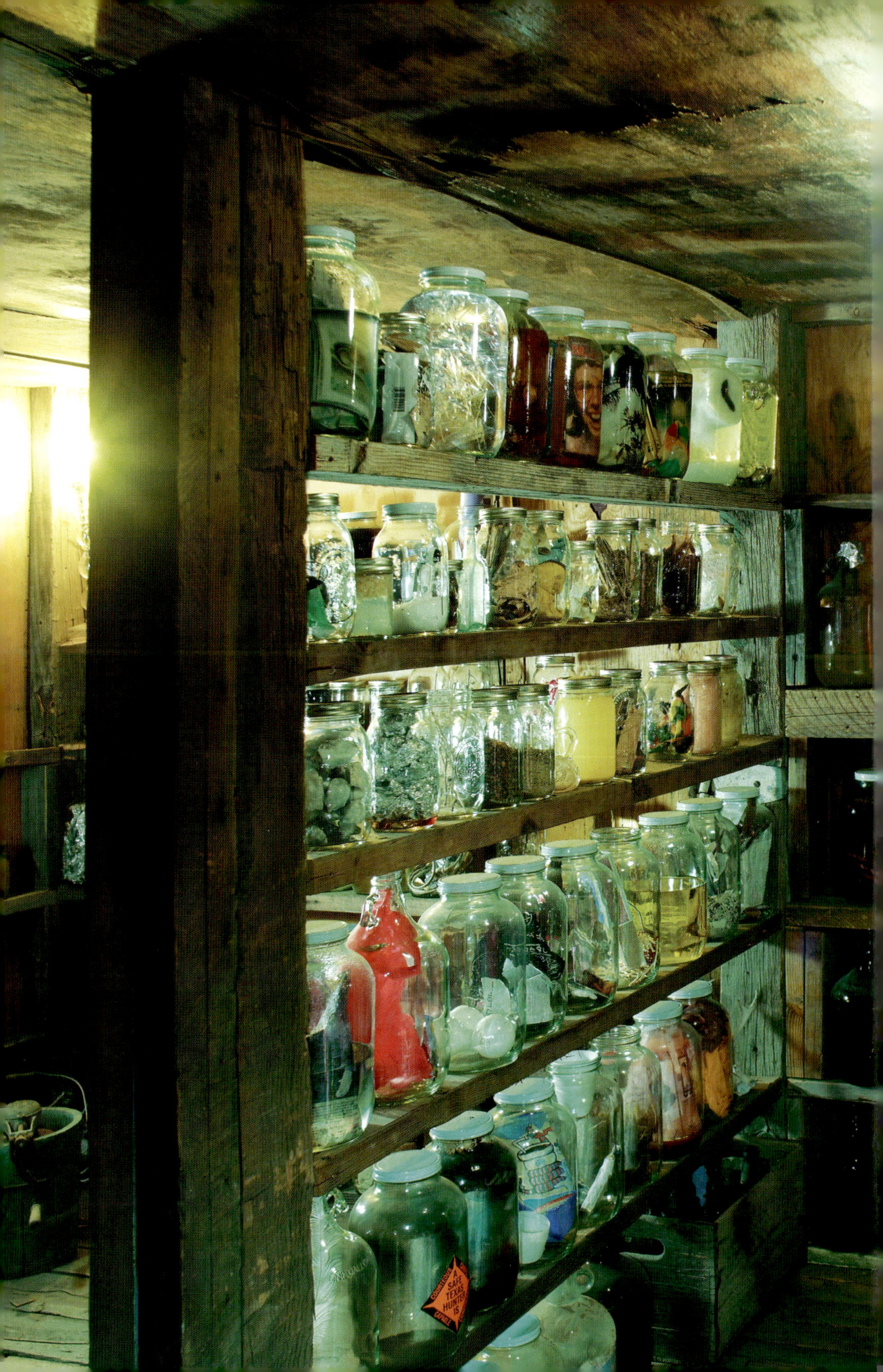
A SAFE TEXAS HUNTER IS
COURTEOUS
CAPABLE

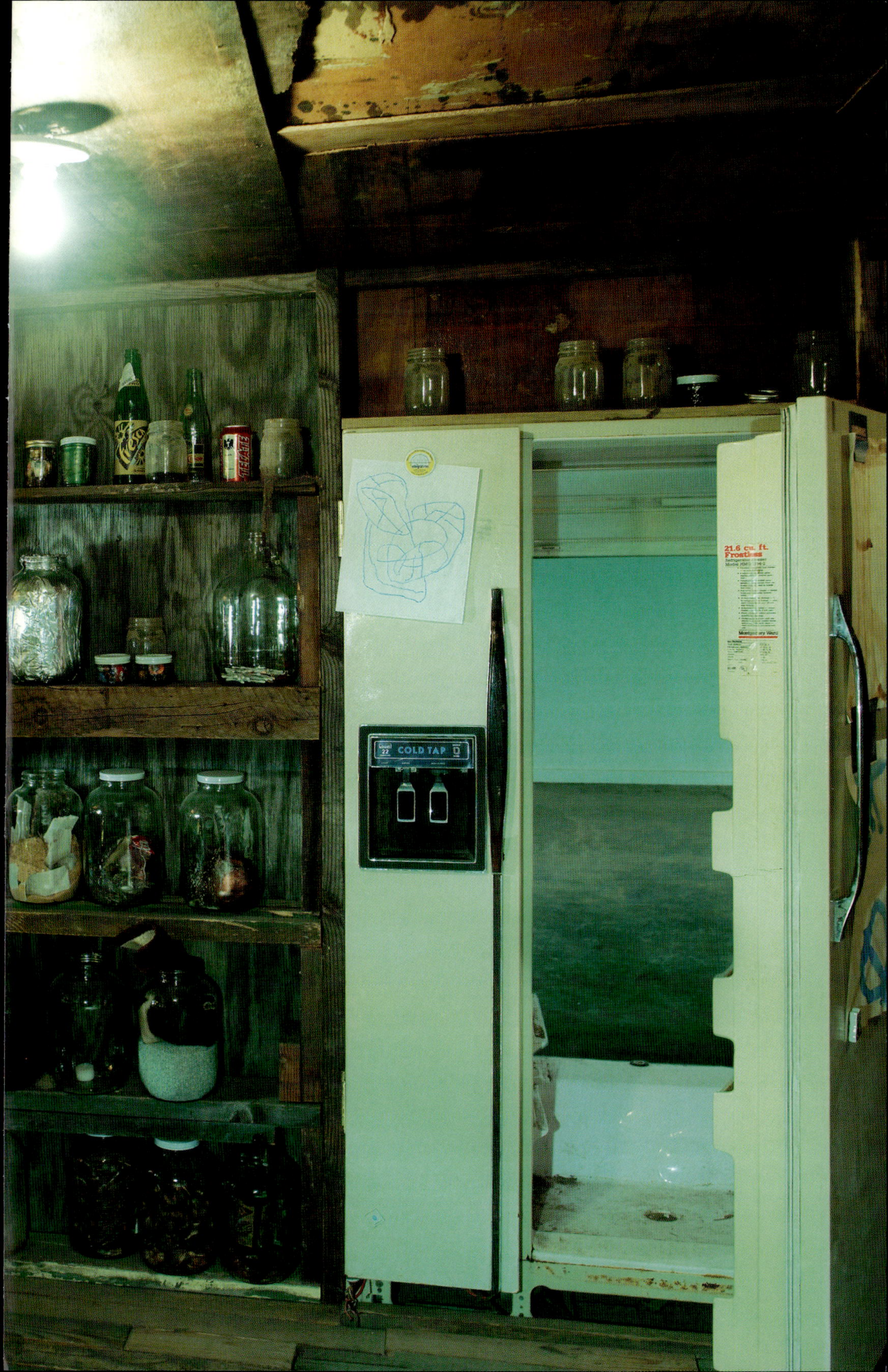
COLD TAP
21.6 cu. ft.
Frostless

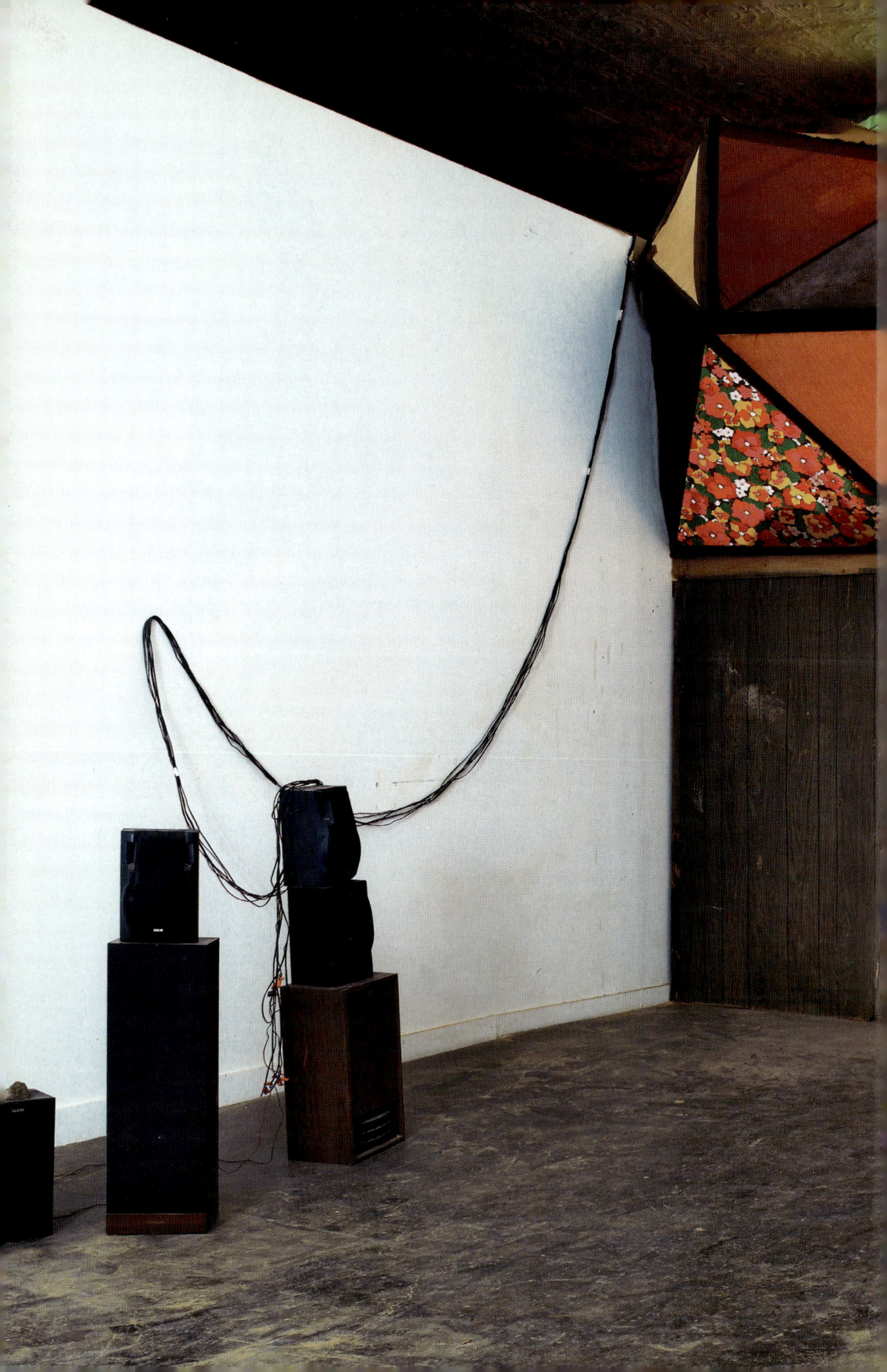

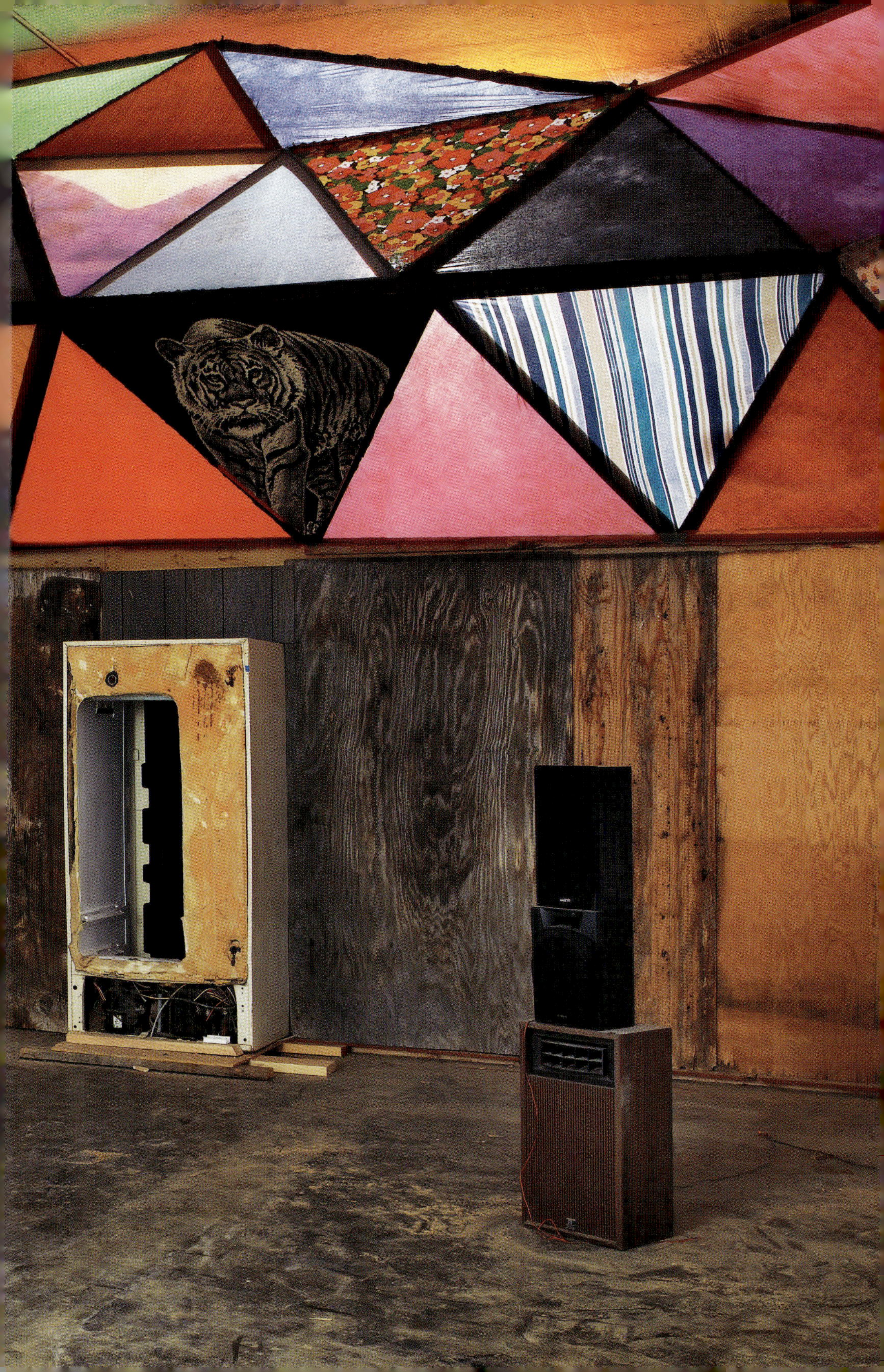

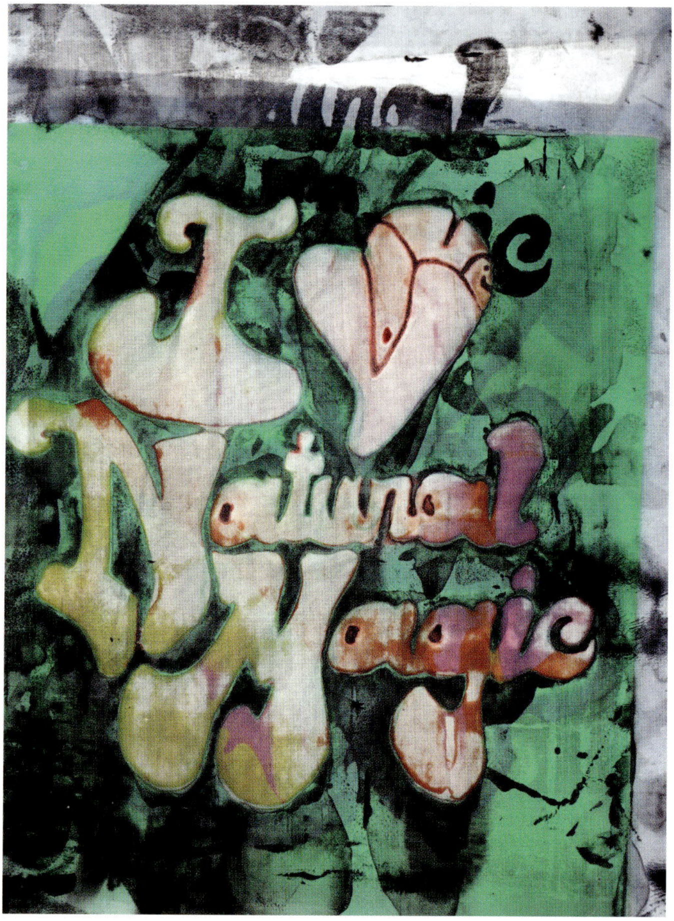
I
Natural
Magic